Beauty Sick

Beauty Sick

How the Cultural Obsession with Appearance Hurts Girls and Women

RENEE ENGELN, PhD

HARPER

An Imprint of HarperCollins*Publishers*

7/9/

BEAUTY SICK. Copyright © 2017 by Renee Engeln. All rights reserved. Printed in the United States of America. No part of this book may be used or reproduced in any manner whatsoever without written permission except in the case of brief quotations embodied in critical articles and reviews. For information, address HarperCollins Publishers, 195 Broadway, New York, NY 10007.

HarperCollins books may be purchased for educational, business, or sales promotional use. For information, please email the Special Markets Department at SPsales@harpercollins.com.

FIRST EDITION

Epigraph on page ix, "The Armpit Song" by Siwan Clark, printed with permission. Excerpt in Chapter 1, "It's Not Your Job" by Caitlyn Siehl, reprinted with permission. Chapter 14, *milk and honey* © 2015 by Rupi Kaur (Andrews McMeel Publishing)

Designed by Bonni Leon-Berman

Library of Congress Cataloging-in-Publication Data has been applied for.

ISBN 978-0-06-246977-9

17 18 19 20 21   LSC   10 9 8 7 6 5 4 3 2 1

To the girls and women who fight, in all the ways they know how, for better tomorrows.

Contents

I sometimes think I could take on the world,

but first . . .

Oh my God, my eyebrows need plucking

And oh my God, my legs need shaving

And my pores need cleansing and my skin needs toning

And my boobs need padding and my hair needs combing

—Siwan Clark, *"The Armpit Song"*

Introduction

I TAUGHT MY FIRST college-level course, "The Psychology of Women," almost twenty years ago. As I got to know the young women who were students in my class, I became both impressed and concerned. These students blew me away with their intelligence and perseverance, their humor, and their consistent openness to engaging with difficult ideas. But some of the worries that burdened these talented women surprised me with their intensity. Of course there were anxieties about things like grades, finding a job, or relationship troubles. But these women also spent an alarming amount of time worrying about their weight, their skin, their clothing, and their hair. One student admitted she missed class one day simply because she felt "too ugly to be seen in public." The other women in the class accepted this claim without surprise, secure in the knowledge that if you're a woman worrying about how you look, you'll always be in good company. After offering the obligatory reassurances that she wasn't ugly, they patted her on the shoulder with gentle understanding.

I recently met up with a friend of mine who is a professor at a small university in the South. As we sat at a local café, catching each other up on our lives, he began telling me a story about the service trips abroad that he leads for college students. A couple of weeks before one of these groups was set to leave for a tropical location, he asked the students to write reflections on whether they were prepared for their trip. Out of the seven women in the group, five wrote that they weren't ready because they had hoped to lose weight before leaving. They seemed more worried about how their bodies might appear than whether they had sufficiently reflected on the work they hoped to do during their trip. Not one man in the group wrote about his body "not being ready." When my friend shared that story with me, my mouth actually dropped open for a moment.

"No," I responded. I didn't want to believe him.

"Yes," he confirmed. "Five out of seven."

"What did you write on their journals?" I asked. "What kind of feedback can you give in response to that?"

He told me he wasn't sure what to write, but eventually settled on reassuring those five women that the culture they were visiting was accepting and nonjudgmental. I doubt that provided much comfort. Even when we travel, we never really leave our own culture far behind, and that's the culture that led these women to write what they did in those journals.

So many young women today are strikingly bold in important areas of their lives, but still crumble in front of the mirror. They fight so hard to be treated with respect, but seem, at least at times, to be willing to trade it all in an instant if they could only remake their physical appearance.

Sometimes I wonder if I and the grown women I know are really so different from that student who skipped class in response to what she saw in the mirror, or the young women who weren't ready to travel because they didn't feel thin enough. Maybe we've never stayed home from work due to a case of the uglies, but how often have we had conversations about our physical flaws, bonding over our displeasure with weight gain or newly appeared wrinkles? How much more time do we spend getting ready for work every day compared to our male colleagues? When one of my most beloved mentors told a group that she wore a scarf every day because her aging neck was "too hideous to be seen," why didn't we question that type of talk? Why are some of my women colleagues still worrying about how many hotness-indicating chili peppers they have on a popular rating website for professors? We may no longer play dress-up and pose in front of the mirror the way many young girls do, but I worry that's only because we've internalized that mirror. We never actually left it behind.

For the past fifteen years, I have dedicated myself to studying girls' and women's struggles with beauty and body image. I often look back to that first class I taught and think about the young woman who wouldn't leave her room one day because she felt too unattractive. I don't believe she's that atypical. I also don't think she was "crazy" or unusually vain. What I do believe is that she was suffering. She was beauty sick.

We should not be surprised at how many women struggle with beauty sickness. We have created a culture that tells women the most important thing they can be is beautiful. Then we pummel them with a standard of beauty they will never meet. After that, when they worry about beauty, we call them superficial. Or even worse, we dismiss their concerns altogether, saying, "Everyone is beautiful in their own way," and admonishing them to accept themselves the way they are.

I wrote this book with the hope that it would provide a path through the miasma of messages we hear about women and beauty. Today's women (and the people who care about them) deserve an honest and challenging assessment of the role beauty plays in their lives, along with scientifically sound advice on how to most effectively push back against a beauty-sick culture.

In addition to scientific research, I will share the stories of a variety of women making their way through beauty sickness. Although these women come from many different walks of life, they are not a representative sample of all women. They are simply women who had stories to tell and were willing to share those stories with me. In particular, they are a relatively privileged group, mostly college-bound or college-educated. Additionally, none are transgender women.

Most interviewees chose to have their names and identifying details changed to protect their privacy and the privacy of other people in their narratives. These women are marked with an asterisk when first introduced. Beyond edits for clarity and changes to protect identities, interviewees' words are their own.

I hope you will find at least one woman in this book whose story speaks to you. No matter our differences, we are all on this road together. The words of others who have walked before us can be powerful guides. The words of those who feel they are walking alone or stumbling can remind us to care for each other.

ONE

This Is Beauty Sickness

1

Will I Be Pretty?

WHEN I TALK to young girls, I often ask that question so many adults ask. *What do you want to be when you grow up?* I love the variety of answers. A teacher. A scientist. An astronaut. A veterinarian. A painter. The president. But no matter what type of life young girls dream of, I know there's a good chance there are two things they really want to be: thin and pretty.

Girls start thinking about their ideal body at a shockingly early age. Thirty-four percent of five-year-old girls engage in deliberate dietary restraint at least "sometimes." Twenty-eight percent of these girls say they want their bodies to look like the women they see in movies and on television.[1] To put this into context, important developmental milestones for five-year-olds include the successful use of a fork and spoon and the ability to count ten or more objects. These are girls who are just learning how to move their bodies around in the world, yet somehow they're already worried about how their bodies look, already seeking to take up less space.

Between ages five and nine, 40 percent of girls say they wish they were thinner.[2] Almost one-third of third-grade girls report they are "always" afraid of becoming fat.[3] These young girls are not worried about their weight because of health concerns. They're worried because they know that being pretty matters for girls, and that in this culture, thinness is a key component of that prettiness.

Leigh*, a bright, delightful seven-year-old girl with a curious disposition, visited my office with her mother, who had agreed to be interviewed for this book. Leigh decided she wanted to be interviewed as well, so I talked with her first. Leigh's mother stayed in the room, but sat slightly behind Leigh so that she would be less likely to influence her responses.

The chair in my office on which Leigh sat was too tall for her, leaving her free to swing her legs as we talked. Leigh's resting facial expression was mildly skeptical, as if she couldn't decide whether the visit was boring, like going to see a doctor, or fun, because she got to play with the toys on the table in my office. Either way, she was a good sport.

"Leigh," I asked, "can you think of what a beautiful woman looks like? Someone who's very pretty? Can you make her up in your head?"

Leigh squinted her eyes a bit and nodded. "She has long, straight hair, and she's wearing a lot of makeup. And high heels. She's thin. Her arms and legs are thin." Leigh's description sounded as though she were reading off a list of specifications for a casting call. After detailing the thinness required for various parts of this imaginary woman's body, Leigh paused. "I'm not sure how big her head is," she said, her brow wrinkled in thought.

The moment was simultaneously sad and charming. Charming because Leigh was so perplexed about how to describe the size of a woman's head. Sad because she already believed that a woman's beauty could be captured via a series of measurements.

I asked Leigh if it matters whether a girl is pretty. "You get more praise and stuff," she told me, barely taking her eyes away from the miniature Rubik's Cube she was manipulating.

Early in many girls' development, the desire to be prettier is already cluttering their thoughts. I'm sure I was no different as a young girl. I remember my grandparents taking me to Cypress Gardens in Florida when I was five years old. In addition to all the lovely flowers you'd expect to see, Cypress Gardens was populated with young, attractive women who had been hired to dress as southern belles and roam the park. They carried parasols and wore frilly, poofy, pastel dresses. I have an album with several photos of young me, clad in shorts and a T-shirt, squinting into the sun, posing next to each of these women. I was too young to wonder why a woman would be hired just to walk around and look pretty, or why there was no male equivalent of the roaming southern belles. I was too young to wonder what it must have felt like to wear one of those crinoline-heavy dresses in the Florida heat, even as my own sweat-soaked hair stuck to my head. I was also too young to ask why all the women were young and white and thin.

Times have changed since that childhood trip to Florida. The cultural obsession with prettiness remains, but the standards are even higher. A relative of mine took her six-year-old daughter to Disney World recently. When the little girl saw Cinderella and Snow White, she complained, "Those aren't real princesses. They're just regular ladies dressed like princesses." She scoffed, "I can tell because their faces are busted."

When I first heard this anecdote, I was confused. I thought she meant their faces literally were broken. Turns up any young person could tell you that *busted* is just a synonym for ugly.

"Where did you learn that word?" asked the little girl's mom.

"YouTube," the girl replied, with a shrug and a grin.

Girls today grow up knowing not just that prettiness is required of women, but that the standard for beauty is near perfection. Even women hired to impersonate princesses leave them thinking, "Meh. I've seen better."

Happily, despite being aware of these princess-level impossible standards, seven-year-old Leigh seems to feel just fine about how she looks. "Leigh," I said, momentarily pulling her attention away from a set of toy magnets, "what if somebody asked, 'What does Leigh look like?' What would you say?"

Leigh scrunches her face up, lets out a long *hmmm*, then answers. "Well, not exactly tall, not exactly short. I'm like the average seven-year-old size, and I have curly red hair and green eyes and today I'm wearing a dark blue dress and light blue shoes."

"That's a good description," I tell Leigh. "What would you say your body looks like?"

Leigh is warmed up now, so there's no pause. "My arms are thin and my legs are really muscly and my trunk is normal."

"Do you like your body?" I ask.

Leigh nods and gives a delightful answer. "I run laps and run around and climb a lot and jump a lot. And I swim and kicking gets my legs good."

"What do you think is more important," I ask Leigh, "if your body can do things or if your body is pretty?"

"Do things." Leigh answers with no hesitation. Leigh's mom smiles from behind her, relief in her eyes.

"Do you think you'll always feel that way?" I ask.

Leigh gets a little quiet. "I'm not sure," she responds.

"I hope so," I say.

"Me too," says Leigh, but she's looking down and her legs have stopped swinging.

I wonder what will happen to Leigh when she enters the rocky territory of adolescence. I hate thinking about the fact that there's a decent chance she will no longer feel so accepting of how she looks. The statistics aren't good. Around 90 percent of young women have no problem naming a body part with which they're unhappy. About

50 percent express what researchers call a "global negative evaluation" of their body.[4] The sense so many teen girls have of not being "good enough" is intimately tied up with the disappointment they feel when they look in the mirror.

Beauty Sick

After researching women's battles with beauty for years, I can confidently tell you that girls and women who struggle to feel at home in their own bodies are not some odd subculture of America. They are not a vanity-struck minority. They are our daughters, our sisters, our students, our friends, our partners, and our loved ones. They are our future leaders. They are sick of wondering if they will ever be beautiful enough. They are beauty sick.

Beauty sickness is what happens when women's emotional energy gets so bound up with what they see in the mirror that it becomes harder for them to see other aspects of their lives. It starts surprisingly early, as soon as young girls are taught that their primary form of currency in this world involves being pleasing to the eyes of others. Although we hear the most about beauty sickness in young women, it's a malaise that affects women of all ages. You can't simply grow out of it. You must break free with deliberate intent and perseverance.

Beauty sickness is fed by a culture that focuses on women's appearance over anything else they might do or say or be. It's reinforced by the images we see and the words we use to describe ourselves and other women. Those who shame women for their appearance feed beauty sickness. Those who praise girls and women only for how they look do the same.

Beauty sickness hurts. It contributes to and finds a ready home in the depression and anxiety that plague so many women. At a practical level, beauty sickness steals women's time, energy, and money, moving us further away from the people we want to be and the lives we want to live. It keeps us facing the mirror instead of facing the world.

Beauty sickness is not a literal illness. You won't see it on an X-ray or in the results of a blood test. But like many other types of illnesses, you can see its widespread and devastating effects. Some of the effects are obvious, like eating disorders and skyrocketing rates of plastic surgery. Others are more subtle, like the distracted hours a girl spends obtaining the perfect selfie to post on social media. Beauty sickness may not be a diagnosis a physician or psychologist would make, but I promise you that any health care practitioner who works with women has seen it. We've all seen it.

If you're a woman, there's a good chance you've felt beauty sickness. If you've ever thought about staying home instead of attending an important event because you didn't think you looked good enough, that was beauty sickness. If you've found yourself distracted during a meeting because you were comparing your body with that of another woman in the room, that's beauty sickness. If you've ever decided not to go swimming with your children because you couldn't imagine facing the world in a bathing suit, that's beauty sickness. If you feel short of time and money, but still spend plenty of both trying to push yourself closer to our culture's beauty ideal, you can blame beauty sickness. If you want to stop worrying about how you look, but keep getting pulled back to the mirror, then you know what beauty sickness feels like.

The signs of beauty sickness are in our thoughts and our behaviors, but this illness also lives in our culture. A beauty-sick culture cares more about an actress's nude selfie than important world events. A beauty-sick culture always, always finds a way to comment on a woman's appearance, no matter how irrelevant it is to the matter at hand. It teaches young girls that learning to apply makeup is a more important skill than learning to do science or math. If you're struggling with beauty sickness, don't imagine it's your own fault. A sick culture makes for sick people.

The Tyranny of the Mirror

After I put out a call for interviewees via social media, I got Artemis's* contact information from her sister. She told me Artemis would be a perfect interviewee. Growing up as a woman in this beauty-sick culture means that just thinking about how you look can be like a sucker punch to the gut. Artemis, a seventeen-year-old high school senior, knows just what that punch feels like. Artemis is of South Asian descent, but because both of her parents were born in the United States, she identifies her ethnicity as "just American." She attends high school in Cincinnati.

I couldn't arrange to meet Artemis in person, so we set up a phone interview. I called her at home on an August afternoon. Artemis apologized for not being able to talk sooner. She'd just returned from a family vacation and was still unpacking. My office air conditioner rattled in the background as we began our conversation.

"What do you think made your sister think you'd be good to interview about body image?" I asked.

Artemis replied using the type of sarcastic tone teenagers wield with such precision. "I think I have a little bit of an idea." She laughed, but didn't directly answer the question. Artemis spent a good portion of our interview standing in front of a full-length mirror with her smartphone pressed to her ear, cataloguing and confirming her many concerns about her appearance. I paced around my office as we talked, trying to understand how this healthy teenage girl had become so terribly unhappy with how she looked.

The first time Artemis remembers being struck with the realization that the shape of her body mattered, she was in seventh grade. Her sister gave her a dress. Short. Sleeveless. Artemis tried it on and stood in front of a mirror.

"I remember, I put it on, and like, I thought that I didn't look good in it. Because I didn't . . . I was too fat for it." Artemis punctuated this memory with a sharp "Ha."

That was one of many ha's in our interview. Artemis would say something tragic, then laugh. She'd make a dire prediction about her own future, and then laugh again. I wasn't always sure where the ironic laughs ended and the real laughs began, if they did at all.

After that moment with the dress, Artemis spent more and more time thinking about how she wasn't as skinny as "everyone else." When she looks back at pictures of her seventh-grade self, she realizes she wasn't heavy by any sane definition, despite being convinced she was "ginormous."

We don't see unvarnished reality when we look in the mirror. Instead, what we see is shaped by years of cultural input, comments from friends and family members, and inner worries. Artemis seemed to know, at some level, that her perceptions were distorted. But she also felt it was unfair of me to ask her to realistically appraise her body size. How was she supposed to know if she really was fat? Maybe she wasn't *technically* fat, but she didn't have to spend long on the Internet to see someone with her body type being criticized for being too heavy.

Though the unrelenting drive for thinness is an important component of beauty sickness, being beauty sick does not mean you have an eating disorder. Eating disorders are deadly and more common than we wish, but the broader epidemic we face is the everyday struggle that girls and women have to feel okay about how they look. Nonetheless, beauty sickness leads many women to dance precipitously around the edges of anorexia or bulimia in pursuit of the perfect body. Artemis's attitudes toward and behaviors regarding weight loss definitely place her in that danger zone. She has started going to bed super early to avoid eating dinner. She's often tired from being hungry, and feels like she doesn't have the energy to do anything.

"A lot of teenage girls who worry so much about their weight can end up heading down the road to an eating disorder. Do you ever worry that will happen to you?" I ask Artemis. Instead of answering my question, Artemis tells me about a friend who had anorexia.

"One of my friends, she's like super skinny. Works out all the time. She looks really good. But she's had an eating disorder multiple times, actually. She's still recovering."

"But you think she looks good, huh?"

Artemis continues enthusiastically, "She's so skinny. She's so toned. I'm, like, 'Damn! I want that body.' Ha-ha."

Artemis has a lot of good things going on in her life. She's doing well in school and thinking about a career in science. She has a close group of friends. But that's not enough, because, she explained, "In the end, the fact that I'm not skinny ruins it all."

This grim statement is a perfect example of beauty sickness at work. It's easy to see where Artemis got the idea that the shape of her body defines her as a person. It's understandable that she feels her body is going to be mercilessly judged every time she sets foot outside her home. A beauty-sick culture never lets women forget that their looks are always up for evaluation by others. Consider just a smattering of pop culture evidence.

- At the 2013 Academy Awards, host Seth MacFarlane opened with a musical number called "We Saw Your Boobs," dedicated entirely

to cataloguing movies in which well-known actresses were top-
less.

- One day after Catherine, Duchess of Cambridge, gave birth to her
first child, she was greeted with a story in the UK tabloid press
about how she could lose the baby weight and "shape up." After
delivering her second child, she was widely criticized for looking
too good after the birth. Writers and commenters suggested she
was harming other mothers because she appeared so soon after
the birth in full makeup, with her hair artfully blown out.

- Heidi Stevens, a popular journalist for the *Chicago Tribune*, wrote
an entire column devoted to all the unsolicited reviews she re-
ceives regarding her *hair*. She shared an email from one reader
who asked, "How could anyone take seriously anything written
by an author whose accompanying picture makes her look like a
tramp, with greasy, matted, uncombed hair?" Remember, she's
not even a broadcast journalist. She writes a *newspaper* column.

- *The New York Times*, the newspaper claiming to decide what's "fit to
print," recently published an article about women seeking curvier
asses ("For Posteriors' Sake"). The article suggested that specific
workouts could sculpt the perfect ass. It was accompanied by a
photo of a young woman in spandex doing a yoga pose over a subway entrance, prominently featured butt high in the air. It's hard
to imagine a similar type of article about men without giggling.

- Blogger Galit Breen wrote a beautiful post about marriage and
lessons learned as she celebrated her twelfth anniversary. She included a candid photo from her wedding. The result was a stream
of hate-filled comments about her weight. One poster suggested
that her husband failed to learn the "don't marry a heifer" lesson.

- When several networks refused to air the 2015 Miss Universe
pageant in protest against remarks made by Donald Trump, the
controversy was widely covered. But news outlets were focused

only on Trump's comments about immigrants. Few were asking an important question. Why, given all the progress women have made over the past few decades, do we still find beauty pageants a completely acceptable form of mainstream entertainment? What is it about our culture that makes the public evaluation of a parade of young women posing in swimwear noncontroversial?

This focus on the appearance of girls and women is so relentless that *Slate* published an article with advice on how *not* to comment on girls' and women's appearance. Perhaps in recognition of our beauty-sick culture, *The Onion* recently published a satirical article describing a beauty industry initiative designed to make women feel self-conscious about a new body part—their palms. Many readers thought the article was real. Women gazed at their palms, wondering.

We See Too Much to "Know Better"

Artemis complained to me that she had gained weight since seventh grade. I told her that was normal. It's what happens during puberty. "Do you think you're overweight now?" I asked.

"I think I am. Ha. I'm not, like, obese, but I've got, like, lumps." Artemis pronounced that last word with a note of disgust.

Lumps? I wasn't sure what Artemis meant. Fat rolls? Cellulite? Or "lovely lady lumps" à la Fergie? Was she talking about breasts? Did she think she was fat because she grew breasts? I pursued the topic a bit more.

Artemis told me she is currently an adult size small or medium. When I noted that this didn't sound big, I was rewarded with an exasperated sigh. I could practically hear Artemis rolling her eyes at how dense I was being, at how much I didn't get it. She tried again to explain the problem. "I've got a little bit of chunk on my side. And my legs are a little chunky. Also, it's mostly in my arms."

Artemis encouraged me to look her up on Facebook and see for myself, so I did. Surprise. She's not ginormous. She looks like a healthy teenage girl. She's got long thick hair and a big smile. I didn't see chunks. Or lumps. Artemis is just over five feet tall, so while it's true that she may never have the stretched-out body of a runway model, I can't imagine any physician who would tell her she needed to lose weight. On top of that, I learned that Artemis is incredibly fit. She exercises one and a half to two hours per day. She runs, she plays tennis, and she plays soccer.

You can probably guess what happened when I told Artemis I didn't think she looked fat. My feedback was immediately dismissed. "Ha!" she said, "definitely fat."

"So if you look in the mirror right now, you're pretty sure you're fat?" I queried.

"I'm pretty sure I'm fat," she confirmed, sounding relieved that I finally seemed to be getting it. Artemis then told me something that I hear over and over from young women. The wording changes a bit, but the sentiment is always the same. Artemis explained, "It's, like, in my brain. I know it's ridiculous that I think I'm fat. But I think I'm fat. I really do." The part of her brain that finds the notion ridiculous doesn't seem to get much airtime.

Artemis's obsession with changing the shape of her body was so overwhelming, I couldn't help but try to challenge her. But her teenage logic foiled my every attempt.

"Could you ever imagine being happy with your body the way it is?" I asked. Artemis responded by launching into a story about the popular Meghan Trainor song "All About That Bass." At the time, I hadn't heard the song, so she described it for me.

"It's her talking about how you don't have to be skinny. Which is actually, like, really motivating. It was a good song. And I was, like, 'Oh, she's actually got a good point. Let's bring booty back! I'm gonna bring booty back!' But then I watched the music video and she was kinda chunky, so I was like, 'Okay, maybe you could pass, but I'd still like to be skinnier than you.'"

Artemis imagines she could wear a new body like armor. She tells me, "I've got, like, a goal, and when I get to my goal I'll be fine. When I go out in public I'll be, like, 'I'm pretty. No one can tear me down!'"

She "stalks skinny girls" on Facebook. She looks through all of their pictures, thinking, "Nope. Not me. I wish." Doing this makes her sad, but she tells me she usually just keeps going until she somehow finds a way to "rip her face away from the computer."

When I began studying this topic, I often got comments about how "stupid" it was for women to want to emulate an unrealistic cultural beauty ideal. One professor even told me that "smart women should know better." It was as though she believed women could magically flip a switch in their heads and just decide the beauty ideal they see hundreds of times a day is irrelevant to their lives. But of course it's not so simple. Few girls and women want to be thin simply for its own sake. Instead, they're taught that looking a certain way is the first step toward getting what they want out of life.

Today's young women face a bewildering set of contradictions. They don't want to *be* Barbie dolls, but still feel they must *look like* Barbie dolls. Many are angry about how women are treated by the media, but they hungrily consume the same media that belittles them. They mock our culture's absurd beauty ideal. They make videos exposing Photoshop tricks. But they can't help wanting to emulate the same images they criticize. They know what they see isn't real, but they still long for it. They download apps on their phones to airbrush their selfies.

Plenty of these women do their best to reject unhealthy beauty ideals. They regularly criticize and question the images they see. But they're also aware of what our culture as a whole says is beautiful.

One of the first studies I conducted when I was a graduate student was in direct response to that professor who told me that disregarding unrealistic beauty ideals should be easy for smart women. I thought her argument was suspect. If it was that easy, we wouldn't see so many women struggling the way they are.

I gave over a hundred young women at Loyola University Chicago the following instructions:

> *Many researchers have studied what the "ideal" woman looks like according to our society's standards. Take a moment to think about what this culture's ideal woman looks like and describe her. Now please take a moment to imagine that you look just like the woman you just described. Think about the ways in which you believe your life would be different if you looked like this woman. How would things change for you?*

The responses were so upsetting that I considered stopping the study early. One young woman told us that if she could just be beautiful, she would finally "be able to focus on her inner abilities and talent." Another said she might "genuinely feel happy most of the time instead of just faking it." A different young woman explained that if she could just look the way our culture wants women to look, she "would never have had an eating disorder, causing stress on everyone around [her] that [she] loved but still hurt." Over 70 percent of the women in this study said they'd be treated better by others if they looked like the beauty ideal.

It doesn't make sense to criticize women for wanting something when every cultural message they receive suggests that thing is the key to happiness. Just like Artemis, many women believe their lives are being profoundly sabotaged by their appearance, and that the only way to salvage things is to change the way they look.

Artemis's story is hardly unique. The countless other women standing in front of their mirrors, poking at the ugly parts, would all understand her predicament. They know what it's like to be beauty sick.

"How often do you think about your body?" I asked Artemis.

She responded quickly. "All day. It sucks. It's just, like, a constant thing. I'll be changing my clothes or something, and then I'll be, like, 'Oh, I should be a little skinnier.' Or I'll be hanging out with my friends and think about how much I would love it if I was skinny like them. It's holding me back from doing lots of things."

Artemis even said she couldn't really start "working on her brain" until she got her body where she wanted it. She laughed as she told me this.

"You're laughing about it, but I wonder if you really think it's funny." I worried Artemis might respond defensively to this bit of analysis, but threw it out there anyway.

"No," Artemis responded, suddenly serious. "It's not funny. It hurts. A lot." Then she laughed again, halfheartedly.

Artemis's insistence that her happiness is completely determined by the size and shape of her body has narrowed her view of her own future. She doesn't even want to think about adulthood and beyond. I suggested that when she was older, she might have things on her mind besides the shape of her body. She found that possibility unlikely. "I try not to think about when I'm old, because I feel like when I'm old I'm inevitably going to be fat and I won't be able to do anything about it. I'm still gonna be sad. I'll be fat *and* I'll be old. A horrible combination."

Artemis imagines a genie offering her three wishes. She would use her wishes for a wholesale bodily transformation. She wants "a nice long neck" and "skinny legs." She doesn't want hips. Not at all. She says she doesn't need hips. She wants a better nose. She wants her hair to be straight. She told me her ideal height and weight and I punched it into an online BMI calculator. It's in the underweight category. It's too thin to be legally hired as a runway model in several countries.

"If I had a magic wand," I told Artemis, "I wouldn't make you skinnier."

"What?" Artemis exclaimed, sounding so angry I briefly forgot I didn't actually have a magic wand. "Yes! You would have to!" Even in the context of this absurd hypothetical scenario, Artemis was upset that I wouldn't help her lose weight.

"No," I continued, "I'd use my magic wand to change the world so that the shape of your body wouldn't matter to you so much anymore."

Artemis wasn't buying it. She said even if the world no longer cared about how skinny she was, she would still care. I pressed her on this logic. Wasn't it something in our culture that gave her the idea that if she were skinnier, she would be happy? Something planted that seed. Couldn't I unplant it with my magic wand? Artemis eventually conceded that perhaps if I could turn back time with my magic wand, if I could erase the link in her mind between thinness and happiness, maybe she could be happy one day. "Awkward!" she declared. Then she laughed sadly.

The Mirror in the Way

The current generation of young women is the most educated we have seen. They are marked by stunning ambition and dogged determination. These are women who scoff at the idea of a glass ceiling. They have bright, wide-open futures. Many have embraced feminism, even if they don't call it feminism. But on their way to changing the world, this generation is wading through a toxic swamp of messages about women's bodies. This swamp feeds an obsession with appearance that starts young and spreads fast. It distracts and depresses.

Women want to *lean in,* but in a culture that teaches them to value their beauty above all else, like Artemis, they're often leaning closer to the mirror instead of closer to their dreams. That mirror becomes a barrier, reminding them that the world will allow them to be powerful, but never so powerful that strangers won't comment on their body shape or insist that they "smile, sweetheart."

A few years ago, I received an email from a Canadian woman who saw a TEDx talk I gave about beauty sickness. She confessed that she almost didn't attend an important fundraising function for a children's welfare organization because she felt she looked too fat in her clothes. After reframing her feelings in terms of beauty sickness, she decided she would "let her looks be a sideshow to who I really am." This choice made a real difference in her life and in the lives of others. "If I hadn't gone," she wrote, "I wouldn't have met some incredible people. I wouldn't have bought raffle tickets to help children living in poverty be able to go to camp this summer. I wouldn't have offered my time to help organize next year's event."

As a culture, we face a terrible loss when an entire group of promising citizens is spending so much time worrying about whether they are beautiful that they risk letting another generation go by without seeing the changes they hoped to see in the world. This obsession with appearance—this beauty sickness—turns women toward the mirror and away from a world that stands to benefit from a refocusing of their passions and efforts. How might women's lives be different if they took the energy and concern aimed at their own appearance and aimed it out at the world instead?

Women have important jobs to do. As poet Caitlyn Siehl expresses so beautifully below, being pretty is not one of them.

> when your little girl
> asks you if she's pretty
> your heart will drop like a wineglass
> on the hardwood floor
> part of you will want to say
> of course you are, don't ever question it
> and the other part
> the part that is clawing at
> you
> will want to grab her by her shoulders
> look straight into the wells of
> her eyes until they echo back to you
> and say
> you do not have to be if you don't want to
> it is not your job

When I originally interviewed Artemis for this book, she chose the pseudonym Violet. A few days later, she emailed and asked me to change it to Artemis. The name Artemis originates from the Greek goddess of the hunt, protector of girls and women. I like to imagine that the switch from Violet to Artemis had something to do with reaching for a more powerful image of who she might be, but it's impossible to know. I imagine Artemis growing up strong and brave, reaching out her arms to protect all the young women suffering the way she does now.

2

Just Like a Woman

AT NORTHWESTERN UNIVERSITY, I've been teaching a course on the psychology of gender for over ten years. It's a fun class for me, in part because of the fascinating examples of gender-related phenomena that flood the media on a daily basis. When Caitlyn Jenner came out as a transgender woman, I paid special attention to the press coverage, thinking it might form the basis of an interesting class assignment. It seemed everyone had something to say about Caitlyn.

Beyond the slew of congratulatory tweets and an equally massive mess of hateful comments, most of what people had to say was about her appearance. Some commentators declared her look on the cover of that now-famous *Vanity Fair* issue "fantastic." Others took the opportunity to ask, "Who do you think has more plastic in their body? Jenner or her stepdaughter, Kim Kardashian?" As soon as Caitlyn identified herself as a woman, nearly every media commentator felt the need to make pronouncements on her body. Conversations about Jenner were dominated by questions about whether she was more or less attractive than other women. As comedian Jon Stewart put it, "Caitlyn, when you were a man, we could talk about your athleticism, your business acumen. But now you're a woman, which means your looks are really the only thing we care about."

In many ways, Caitlyn Jenner is living a life most women would find foreign. Few of us have a team of makeup artists and hair stylists, a beach house in Malibu, or a reality show. But there's something all women have in common with Jenner. We know what it's like to live in a world where our appearance is treated as paramount, where strangers comment on our choice of clothing or our body shape. We know how it feels when how we look seems so often to matter more than our character or our actions. And we are no different than Caitlyn when it comes to the knowledge that no matter what we do, we'll never look good enough to please everyone.

We can't talk about beauty sickness unless we're willing to acknowledge the role of gender. Some men suffer from appearance worries, many quite significantly. But men are not the focus of this book. The depth and breadth of the influence of beauty concerns on women's lives means that, on average, looking in the mirror is a substantially different experience for women than it is for men. Beauty sickness may not be exclusive to women, but it is overwhelmingly a women's issue.

Consider a recent controversy over a set of children's onesies on sale at an NYU bookstore. A photo an employee snapped of the display went viral. On the left, a purple onesie intended for little girls read, "I hate my thighs." On the right, for boys, we saw a blue and yellow onesie complete with a cape. It read, "I'm super." The makers of the "I hate my thighs" onesie claimed it was intended to be ironic—after all, regardless of gender, healthy babies have big, squishy thighs. But it struck many as no coincidence that the body-loathing onesie was intended for little girls. The training to feel disgust toward your own body is part of growing into a woman, and it starts early.

Gabrielle*, a thirty-three-year-old social worker and mother to an eleven-year-old daughter, had an unusual experience when it came to childhood lessons about beauty. She was born in Portugal and spent most of her childhood there, then bounced around Western Europe before settling in Florida in her twenties. We met in a noisy coffee shop. It was a sunny afternoon, and the shop was packed with tourists and office workers escaping for a midday break. Gabrielle arrived dressed casually in jeans and a fitted T-shirt, but her face was fully made up and her long dark hair was flawlessly arranged in the kind of waves you usually see only in magazines. Gabrielle seemed not to notice, but she turned heads when she walked through the shop. She has that look of a celebrity you can't quite place.

I hadn't anticipated the near-constant drone of a blender in the background, so Gabrielle and I had to huddle in a corner and lean close in order to be heard. It felt a bit conspiratorial, especially when Gabrielle pulled out a photo album and said, "I brought something to show you. Otherwise you would never believe me."

Growing up, Gabrielle received vastly different messages from her mother and father about beauty and gender. Her mother was dedicated to shattering the link between beauty pressures and womanhood, or at the very least ignoring it. Her father did the opposite, encouraging Gabrielle to see beauty as a key source of power for women.

Gabrielle's mother, a successful medical researcher, found makeup and other such "girliness" a waste of time. The more girlie something was, the more likely Gabrielle's mother was to ban it. Gabrielle wasn't allowed to pierce her ears or play with Barbie dolls or grow her hair out. Her mom emphasized practicality above all else, keeping her own hair short, wearing glasses instead of contacts, and eschewing makeup and heels. She did her best to mold Gabrielle in her image, hoping Gabrielle would grow up to believe that "beauty was on the inside." She wanted desperately to have a daughter who didn't worry about how she looked, and the only way she knew how to do that was to try to make Gabrielle more of a boy and less of a girl.

When Gabrielle first opened the photo album she brought to our meeting, it was to show me a picture of her at five years old. I never would have recognized her. She looked like a little boy. Her hair was in a bowl cut and she wore a red T-shirt and overalls. I thought the picture was pretty cute, but Gabrielle shook her head as she pointed to her young self, making it clear to me that *she* did not find it cute. She remembers that period of her childhood as being marked by pleas for dresses instead of overalls. At five, she was old enough to know she looked like a little boy, and she didn't like it. Once young Gabrielle was out with her dad. She was crying, tired and frustrated because she wanted to go home. She remembers one of her father's friends trying to comfort her, saying, "Boys don't cry! Only girls cry. Boys aren't supposed to cry." Gabrielle cried more.

Gabrielle would beg her mom not to cut her hair. "But it dries faster that way," her mom would reply. Young Gabrielle longed for a more conventionally feminine appearance. She even told me that she felt she could relate a bit to how transgender individuals must feel, because when she was young, she knew people saw her one way, but she felt completely different on the inside. Gabrielle promised her mom that when she grew up, she would "wear earrings and high heels and dresses every single day, and have hair down to my knees."

Complicating all of this was the fact that Gabrielle's father was passing on a very different set of messages, telling Gabrielle that beautiful women "could get whatever they wanted." He described women's beauty as a power to be nurtured and used at will. Throughout her childhood, Gabrielle's father reminded her that "you never know who you're going to meet."

"You might bump into the president!" he would explain. "Or there could be a television camera somewhere. You might end up on TV without realizing it!"

Gabrielle's mother told her appearances were superficial and her father told her nothing mattered more than looking good. Whom did she believe? Her father. Hands down. She described him as "planting a seed that grew." It seemed no coincidence that Gabrielle arrived for our interview wearing camera-ready makeup. After all, you never know whom you might run into.

Once Gabrielle neared adolescence and was finally allowed the freedom to start cultivating a more feminine appearance, strangers commented on how pretty she was. She felt "greedy" for that type of attention. She'd never had it before and couldn't get enough of it. But Gabrielle paid a price for entering that world. By age thirteen, she was in the throes of beauty sickness. Puberty widened her hips, and she fought back by cutting her calories to dangerously low levels, sometimes eating nothing but an apple and a banana all day. She'd skip lunch and use the money to buy makeup instead. She just wanted to be "like the magazines." She read *CosmoGirl* as if it were the Bible. Gabrielle was terrified of getting fat. Her body was becoming curvy, but the girls in the magazines were petite and narrow. The more Gabrielle tried to follow the path set for women, the worse she felt. Joining traditionally feminine culture left her feeling more pressure to meet beauty standards at the same time that she felt less *able* to meet them.

Women Feel Worse About How They Look Than Men Do

Psychology has a long and ugly history of misconstruing and exaggerating findings of gender differences. I'm not a fan of the unfortunate metaphor that men are from Mars and women are from Venus. Thinking like that causes us to ignore the fact that many gender differences are smaller and less consequential than we imagine. However, when it comes to beauty sickness, the gender gap is real and big. And it's reliable, showing up time and time again in cultures all around the world.

Think about it this way. If I told you that someone in the United States was about to go under the knife in the hopes of reshaping their appearance, you could bet it was a woman and be correct around 90 percent of the time. Internationally, 85 to 90 percent of surgical and nonsurgical cosmetic procedures are performed on women. If you heard that a young person was suffering from anorexia or bulimia and guessed it was a young woman, you'd be right nine times out of ten. Eating disorders and plastic surgery are both complex issues, but we can't ignore gender gaps like these. They force us to acknowledge that men's and women's experiences in this culture are vastly different in important ways. Men and women are living systematically different lives when it comes to beauty sickness.

Women talk about how they look more than men do, they think about how they look more than men do, and they're more likely than men to engage in behaviors to alter or improve their appearance. There's a reason no one in this culture would be surprised to overhear a woman saying, "I feel so fat and ugly today." We accept this type of unhappiness as part of *being* a woman. Over thirty years ago, researchers coined the phrase *normative discontent* to describe this phenomenon.[1] The term suggests that we've gotten to a place where it's considered normal for girls or women to be deeply disappointed when they look in the mirror. It's the girl version of "boys will be boys."

Women's struggles to feel comfortable in their own bodies should not be treated as simply a fact of life. They are not the result of some inevitable force. We are not born this way. We know where this pain comes from. This pain starts when girls and women learn the lesson that teenage Gabrielle learned. They learn that the most important asset they have is their physical beauty. We don't teach boys and men this same lesson. The end result is that women often rightly feel they're moving around in a different world than men are, a world where they can't escape the mirror. And when they do look in the mirror, women are much less likely to like what they see.

There's a statistic that frequently shows up online if you search for anything having to do with women's body concerns. "Fifty-four percent of women would rather be hit by a truck than be fat." It took me over an hour and the help of a fact-checking friend to find the original source of this oft-cited finding. It's from *Esquire* magazine, back in 1994. It's certainly not based on a careful scientific study, but there's still something compelling about it. I once shared that data point with the women in one of my classes, imagining they would be shocked and appalled. Instead, they asked questions about the truck accident. How big is the truck? What kind of truck? How fast is it going? How much would it hurt? Their fear of losing the battle for the ideal body was so potent that pain and injury were less threatening. Teenage Gabrielle would have chosen the truck too, stepping right in front of it. She feared that "nobody would ever like her" if she gained weight, and she was willing to do whatever she could to avoid that fate.

Once Gabrielle was given free rein to start work on the project of her appearance, it became an all-consuming job. When she joined the ranks of girls and women yearning to make the image in the mirror match the images in the magazines, she opened the door to waves of insecurity that still wash over her today. Gabrielle didn't want to look like a boy, but she didn't intend to sign up for so many unhappy hours in front of the mirror. She unwittingly traded one painful situation for another.

An analysis of over two hundred published studies on body image found that the gender difference in body dissatisfaction has been around for decades, and seems to be getting bigger over time.[2] A recent World Health Organization report called "Growing

Up Unequal"[3] surveyed over 200,000 young people in forty-two different countries. I was saddened but not surprised to read that among fifteen-year-olds, in every country surveyed, girls were more likely than boys to report that they were "too fat." This finding emerged despite the fact that boys were more likely to actually *be* overweight. Being a girl was a better predictor of whether a young person felt "too fat" than actual body size was.

Unfortunately, there's little evidence that girls grow out of body dissatisfaction as they turn into women. Instead, just as Gabrielle's story exemplifies, puberty often packs an unexpected punch for girls. Because going through puberty typically increases body fat in girls, especially in the hip and thigh areas, it has an ironic effect. This physical maturation process that turns girls into women leaves them with bodies that look *less* like our culture's super-thin body ideal for women. The psychological outcomes of the bodily changes associated with puberty are a key point of gender divergence. One study of over four hundred children in the southwestern United States found that boys' body satisfaction actually improves as a result of puberty.[4] Boys tend to gain muscle mass and move closer to their ideal body at the same time that girls are moving further away.

Consider fashion designers who hire girls as young as fourteen years old to model women's clothing on the runway. At a certain point, our body ideal became so removed from what most women look like that it was easier to find the ideal in young girls than grown women. You might imagine that a recent reemphasis on highly curvy women's bodies (think Kim Kardashian, Christina Hendricks, or Beyoncé) would have changed this trend, but there's no evidence of that happening anytime soon.

Although body dissatisfaction is a major driver of beauty sickness, women's appearance woes aren't limited to body shape and size. Even if you look at general appearance satisfaction, taking into

account things like one's face, skin, and hair, women are still struggling compared to men. One survey of over 50,000 adults found that
women were less satisfied overall with how they looked compared
to men.[5] This result held from ages eighteen to sixty-five, again
suggesting that you can't simply grow out of these concerns. Inter-
estingly, the possibility that women feel less attractive than men
because they actually *are* less attractive than men isn't borne out by
the data. Despite the fact that women may feel uglier than men do,
when observers rate the attractiveness of images of strangers, they
consistently give women higher attractiveness scores than men.

Psychologists often talk about self-serving biases, which facil-
itate the general human tendency to think we're better at things
than we actually are. For example, most people will tell you they are
an "above average" driver, even though this is statistically impossible. Likewise, both men and women tend to overestimate their own
intelligence. But when it comes to evaluating one's own physical
appearance, the self-serving bias shows up in men but disappears
among women. In an article on these "narcissistic illusions," researchers studying college students in Texas found that men overestimated their attractiveness and women did the opposite.[6]

The gender difference in the experience of one's body extends
beyond simple satisfaction or dissatisfaction. When researchers
at the University of Sussex interviewed dozens of British men and
women,[7] they found that women tended to view their bodies in a
fragmented way. They described their body parts as a series of dis-
appointments, punctuated with rare parts that were "just okay."
Most women seem to have a catalogue of appearance complaints
at the ready. Stomach too wobbly, thigh gap nonexistent, skin un-
even, hair not shiny enough. Each part stands on its own, ready to
be picked apart.

When you ask men how they feel about their bodies, they tend to take a more holistic approach. They experience their bodies as one unit, not as a series of pull-apart components that need to be altered or fixed. Perhaps most important, men think much more about their bodies' capabilities. Every single man interviewed in the study referenced above talked about his body in terms of what it could *do*. Not *one* woman did. Perhaps without even realizing it, these women had internalized the message that women's bodies are for looking at, not for doing.

Men and women differ also in the emotional experience of being in their bodies. In one study out of Duke University, men and women were asked to try on a bathing suit and stand in front of a mirror. They were alone—no one was there to see what they looked like. Whereas men said they felt "silly" in the bathing suit, women's emotional experiences were much more intense. Wearing a bathing suit in front of a mirror left the women feeling disgusted and angry, even revolted.[8] How can you have respect for yourself as a human being if you're disgusted by such an important part of your own humanity? How you feel about your body's appearance is inextricably linked to how you feel about yourself, and this link between self-esteem and body esteem is stronger in women than men.[9] This is why, when people want to inflict emotional damage on a woman, they so often make a strike at physical appearance. Even Leigh, the seven-year-old girl you met in chapter 1, knows this. When I asked her, "Leigh, what's the meanest thing you could say to a girl?" she answered: "You're fat and you're ugly."

Too often, for women, the body becomes an enemy. It becomes something that must be tamed into submission. Gabrielle tried to tame her body with starvation diets and layers of cosmetics. Other women employ expensive beauty treatments, cosmetic surgery, or obsessive exercise routines—anything just to move one's body closer to a place that might finally feel good enough.

Beauty Can Be Powerful for Women, but It's a Weak and Temporary Power

When young Gabrielle felt fat and ugly, she'd go to her parents for comfort. Gabrielle explained, "Any time I was feeling insecure, like, if I didn't look good or had acne or something, first I would go to my mom. She'd say, 'That doesn't matter! None of that matters.'"

Unsatisfied, Gabrielle would seek out her dad. "You're the most beautiful girl in the world!" he'd say. Gabrielle didn't believe either one of them.

By the time she had reached age fifteen, things had shifted for Gabrielle. She settled into her new adult body and found that adult men appreciated it. They were drawn to her then, just as they are today. "I noticed that people sort of came to me. I didn't have to do any work whatsoever. People just gave me this power because I looked a certain way. It was undeniable. And once I had that power, I didn't want to lose it."

During college, Gabrielle took on a series of what she called "appearance-related" jobs. Minor modeling gigs, being a "hostess" at trade shows. "It paid more than working at Burger King," she explained. And it fed the need to know that others found her beautiful. Not surprisingly, Gabrielle's mom was appalled by those jobs.

"It's mind-numbing, like you're just a mannequin. You're being objectified," she worried.

"What's wrong with that?" Gabrielle demanded.

I didn't get to meet Gabrielle's mom, but I think I know what was bothering her when it came to the jobs Gabrielle was taking on. I'm still mortified about one of the ways I paid my rent and tuition when I was in college. It was nothing too sordid, but in addition to bartending and waitressing at a restaurant on campus, I occasionally did what was politely referred to as "PR work" for a local distributor.

It entailed wearing a beer-branded tank dress, absurdly high heels, and standing for a couple of hours at a time while men at least twice my age waited to have a Polaroid picture taken with me and two other college women who held the same job. None of us took the gig very seriously. We were always in some small town miles from campus, unlikely to be spotted by anyone we knew. We were also, thankfully, years away from digital photography and social media. We cashed our checks and moved on with our lives. I lost that tank dress years ago when someone borrowed it for a Halloween costume and never returned it.

I blush when I think about how out of line that job was with the values I have today. But it's more than just embarrassment. I feel anger that I didn't have a framework for questioning what I was doing. It was my own variant of the Cypress Garden southern belles. The late adolescent version of me never thought twice about the implications of those types of jobs. People weren't really talking about these issues when I was young. Even if someone had pushed me about the choice to work that job, I imagine I would have responded the same way teenage Gabrielle did. "What's wrong with it?" I would have asked. But I did feel vaguely uncomfortable, smiling for a camera in my skintight dress. I just didn't have the words to explain why.

Today I have the words. When we pay women to be looked at, we send an important message about what women's bodies are for. We implicitly acknowledge that they're decorative. They're passive. At heart, I'm a pretty nerdy bookworm. But when I was in that dress, my intellect disappeared. It wasn't a part of me. It wasn't relevant. Beyond that, my body became something that existed for other people. It was static, captured again and again in pictures. I was, during those periods, a well-paid statue who knew how to smile. I wasn't the complete human being I wanted to be. There is no real power in that type of passivity.

At a basic level, even our language reflects the idea that women's bodies are passive, whereas men's are active. The roots of the word *handsome*, men's version of beautiful, point to being *handy*. As the *Oxford English Dictionary* puts it, the original sense of handsome was "suitable, apt, clever." *Beautiful*, on the other hand, is generally defined with a focus on pleasing the senses or being ornamental.

We see this linguistic difference borne out in men's and women's daily lives. Though men absolutely face appearance pressures too, they live in a world where their competence is generally prized more than how they look. In essence, they can trade a focus on appearance for a focus on being *good at* something. Their success in a given domain can provide a safe retreat from appearance pressures.

Women have no such safe place. It doesn't matter how good a woman is at her job, we will still talk about how she looks and demand more from her in terms of her appearance than we would from a man doing her same job. We saw this happen when then Secretary of State Hillary Clinton was roundly criticized by news media for appearing in public wearing glasses and little makeup. The fact that she wore a hair scrunchie became a national topic of conversation. President Obama ran into trouble a few years ago when he introduced Kamala Harris, attorney general of California, as "by far the best-looking attorney general in the country." He first praised her skill and thoughtfulness, but the follow-up comment on her looks demonstrated how no show of competence can keep us from reminding women how important it is that they be pleasing to the eye.

And it's not just men who focus on the appearance of high-profile women over their deeds—women are just as adept at this type of behavior. Not long after criticizing Donald Trump for insulting Carly Fiorina's face, *The View* cohost Joy Behar did the same thing herself—saying Fiorina's face looked "like a Halloween mask." Women don't just hold themselves to impossibly high standards— they often do the same to other women. In a recent *New York Times* article about women's "shapewear" (e.g., Spanx), a successful attorney was quoted as saying, "No woman should ever leave the house without Lycra on her thighs. I don't want to see my own cellulite, so why would I want to see yours?" If your own body disgust gets strong enough, don't be surprised if you sometimes find yourself aiming it outward at other women.

For women, we learn, beauty is the power that matters most. Today Gabrielle still works hard to keep that power, but feels shame over doing so. She calls it "superficial" to worry so much about how she looks, but she also knows that people treat her differently when she looks her best. They're nicer, more willing to help. She describes her struggle as a duality. "I want to be enlightened. But my ego's in the way. I know I shouldn't care about beauty, but I do. I don't want to care, but I can't let it go." She points to her body, lifts her hair, and says, "I know I'll get old and lose all of this, but I want to keep it for as long as I can." You can hear both of Gabrielle's parents' voices when she describes this internal battle. "It feels good when you look good. But looks shouldn't matter. But they *do* matter."

It's not unusual to see articles or books exhorting women to take advantage of the power their beauty offers. It might not be fair that people care so much about how women look, these authors claim, but if this is the only kind of power our culture is really willing to give you, you might as well use it. It's no secret that beauty is a kind of currency for women. It does offer a type of power over other people. But let's be honest about what kind of power this is.

First, it's a type of power that's almost impossible to earn if you weren't born into it. One of my grandfather's favorite things to say to people was "Never be too proud of your youth or your beauty. You did nothing to earn them and you can do nothing to keep them." We forget, too often, that beauty is not democratic. It's not handed out to those who somehow most deserve it. For that reason, there will always be an inherent injustice in the power that beauty gives to women.

On top of that, the power beauty gives resides on unstable ground. It's power that exists only if others are there to acknowledge it. It's never really your own power, because there's always someone else in charge. Even worse, it's power with a strikingly strict expiration date, because the link between youth and beauty is near universal. (It's power you don't get to keep. It's power that expires sometime after your thirties—maybe your forties if you've got a personal trainer, Botox, and a stylist. Women should become *more* powerful with age, as they gain valuable skills, experience, and wisdom. If we tie our power to our beauty, we risk letting it trail away with our youth.) It's a grotesque kind of power that begins to disappear just as a woman starts to find her footing in the world. It's a twisted power that makes women terrified of "showing their age," while men can rest comfortably in the privilege of looking more "distinguished" as they get older.

Gabrielle has thought a lot about the unearned power of beauty. She wonders how she will negotiate a future without that power to compel all eyes in the room. Will she have any power left at all when her body ages away from the youthful ideal? Can she teach her daughter that there is no shame in a woman's aging body?

"Sometimes I see these women. You see this face that has a ton of plastic surgery and her breasts are so big. And my first reaction is to think, 'Who does she think she is? Why did she do that to herself? She'd probably look better if she didn't do any of that.' But then it's really scary because at the same time, I recognize something in me that could do that. A little nip and tuck here and there, then I'm getting carried away. So I try to be aware of that and stay grounded. But it's hard."

It's difficult work for Gabrielle, remembering what her father taught her about beauty. She rarely leaves the house without first applying a pretty serious amount of makeup. There was a time when she wouldn't be seen without makeup, even when in her own home. She told me that in the past year or two, she's gotten "brave," because she has occasionally run errands without first applying cosmetics. But Gabrielle still spends more money on makeup than anything else. It takes her at least thirty minutes a day to do her makeup, more if she's "going out." I asked her to add up all that time and think about whether she would like to do something else with it. Gabrielle is going to school in the evenings to earn a master's degree and she has a "tween" daughter to keep track of, so she's pretty busy. First Gabrielle laughed and said, "Sleep! I need more sleep." Then she shared something her mom always said.

"She'd say, 'You don't have time for everything.' She'd tell me that if you were focused on how you look, you weren't focused on your internal qualities. You would be spending too much time focused on the wrong things." Gabrielle looked away as she thought more about what her mom had said. She sees some truth in it now. "If women didn't worry so much about how they look, they could use that time and mental energy to work on something, you know? To be more concerned about status in the workforce or competing for equal salary or advancing their education."

Gabrielle is still trying to have it all, but she's tired. The beauty work wears her out.

Gabrielle tries not to give her eleven-year-old daughter messages about beauty. Now she sees magazines like *CosmoGirl*, the same magazines that formed her own ideas about women and beauty, and thinks they're "disgusting." Unlike her own mom, Gabrielle doesn't deliberately keep "girlie" things from her daughter, but unlike her father, she doesn't encourage them either. She wants her daughter to be "whatever she wants to be." But Gabrielle hears her daughter say things like "Mom takes so long getting ready!" and "Mom is makeup obsessed!"

How can she keep her daughter away from the pit of beauty sickness if she hasn't pulled herself entirely out yet? Gabrielle wonders if there's anything a parent can do to protect a daughter. "It starts at such a young age, that little girls are told they have to be pretty. My mom didn't do it, but the rest of society did, and that was enough,"

Beauty Sickness Is a Barrier to Gender Equality

A friend of mine works as a computer programmer and recently attended a convention for his job. It was a typical convention in most ways. Different vendors set up booths and employed a variety of tactics to gain the attention of attendees. Raffles. Booze. Free food. This convention was in Chicago, where the Blackhawks hockey team is a pretty big deal. My friend learned that his company was paying to have the Blackhawks' Ice Crew come to their booth, in hopes of drawing a crowd. Feel free to laugh at this, but I assumed the Ice Crew consisted of the people who drove the Zambonis. I thought that was pretty cool.

It ends up the Ice Crew is the hockey equivalent of the Dallas Cowboys Cheerleaders. They skate around in tight uniforms and get the crowds riled up. I was speechless when I learned this. I can't imagine how this must have felt for the handful of women who work at my friend's company. Or for the few other women attending the male-dominated convention. This particular anecdote is not a fluke. Microsoft and Xbox recently got into hot water for hiring dancers dressed as scantily clad schoolgirls (think white bras and short little plaid skirts) to populate a conference after party.

Around 90 percent of software engineers are men. Most high-profile software companies claim to be working to shift this gender imbalance, and several impressive nonprofits are currently dedicated to encouraging girls to get more interested in coding. Is this the world those girls will have to enter? Can they be taken seriously by their male colleagues if their employer is paying women to work as decorations at professional conventions?

Today's young women were raised to believe they could be anything, yet they are still haunted by the need to be pretty above all else. It has been over three decades since young women passed young men in college degrees earned. As confident as today's young women may seem in the classroom or the office, too many are made anxious or depressed by the feeling that their appearance is under constant examination, because in this beauty-sick culture, it probably is.

When women are distracted from their dreams by appearance worries, they risk letting the leadership roles they craved for themselves slip by. They risk losing their voices to a barrage of imagery that slams them with the message that skill and hard work will never be more than icing on the cake of their appearance. Our world needs these young women. We need them to be strong and healthy, leading the way toward a future they are uniquely suited to build. This desperation to be beautiful is not just a threat to women's mental and physical health. It is a roadblock on the path to gender equality. We can't keep pulling ourselves over that roadblock one at a time. We need to take a closer look at its foundation so we can figure out how to dismantle it together.

3

I, Object

WHEN I BEGAN graduate studies in psychology, I didn't plan on becoming a professor. I assumed I would become a clinician, perhaps working in a hospital or building up a private practice. I hoped to focus on women with eating disorders and those who survived trauma. I was only twenty-three when I first started training as a therapist, seeing clients at a university clinic. At that young age, I undoubtedly lacked perspective in some pretty notable ways, but I worked hard to understand the young women I was seeing in therapy.

The very first client I had was a college student who couldn't shake the feeling that she was always falling short. She told me that her trips to our therapy sessions were often difficult for her, because she felt people were "looking at her" every time she walked across the busy campus. When I shared that comment with my supervisor, he encouraged me to see it as a sign of paranoia. "Perhaps she is becoming psychotic?" he offered. He never considered the possibility that people *were* looking at her when she walked across campus. Instead, my client's experiences were understood to be part of some disorder. Her feeling of being looked at must have been the result of faulty wiring somewhere underneath that mess of curls she would twist around her finger when we met in our weekly sessions.

When I look back, that young woman's sense of being observed by others and my supervisor's reaction to it seem emblematic of a bigger issue. There are many women who suffer as a result of feeling chronically gazed at. You can't just brush off that feeling; it matters. It's the core of beauty sickness. We can blame neurotransmitters, hormones, dysfunctional attachment styles, or problematic thinking patterns all we want, but if we focus only on those types of targets when trying to understand women's experiences, we're missing the bigger picture. Any woman who finds herself beauty sick had a lot of outside help getting there. She was most certainly pushed and prodded by a set of cultural factors largely outside of her control.

Even though you may be the one hurting, the source of the problem isn't necessarily *in you*. It's perfectly reasonable to consider the idea that the source is often our beauty-sick culture, a culture that makes women feel so looked at, so often.

Eventually, I left the therapy room to focus my attention on teaching and research. But I often think back to my first client. I wonder how she's feeling today. I wonder if she still feels "looked at." My guess is that she does, and if so, she's in good company with other women all around the world.

If we want to free ourselves from the vise grip of the mirror, we need to take a broad look at what's happening to girls and women. If we want to know who's at fault for Artemis's obsession with losing weight and Gabrielle's fear of aging, or if we want to know why it is so hard for so many of us to feel comfortable in our own bodies, we must look to our culture for answers. It's easy just to blame "the media" for promoting an unattainable beauty ideal, but the problem isn't that simple. In a different type of world, that beauty ideal for women wouldn't *matter* so much. That beauty ideal wouldn't wield such destructive power if there wasn't something more systemic happening in our culture. But before we explore this larger cultural story, let me tell you about Erin*.

ERIN IS A WHITE TWENTY-SIX-YEAR-OLD New Yorker. She's an art student by day and a server at a busy restaurant in the evenings. We met at that restaurant in the quiet period before her shift started and the tables filled. If you take one look at Erin, there's no question that she's an artist. She plays with visual cues and mixes masculinity and femininity in unexpected ways: one day a hot-pink Mohawk and yoga pants, the next day a brightly patterned headscarf, cat-eye glasses, and a polka-dot dress. She does things with makeup that I can't even begin to describe, often looking like she just stepped off the set of a costume drama. Erin loves the *idea* of beauty. She's an artist, after all. But she's simultaneously angry at the way the world seems to define her by her appearance.

There are a lot of things I love about Erin. She is full of surprises, speaking gently yet swearing like a sailor. Erin has a keen eye for injustice, and when it comes to people being treated unfairly, she's always up for a fight. Having a conversation with Erin leaves you feeling taken care of, like you have a motherly warrior on your side.

Erin began by telling me about her identical twin, Meredith*. Identical twins are fascinating. It's really hard *not* to look at them. What that meant for Erin is that she and her sister received twice as much appearance-related attention. "Pretty twins" was often the first thing people said when they met her and her sister. On a superficial level, this sounds nice. You might even guess that Erin grew up having a lot of confidence and feeling pretty. But that's not quite what happened.

An identical twin has the potential to act as a mirror. When you look at your twin, you'll always be, in some ways, looking at yourself. I tried to get a sense of what that was like for Erin as a child. "Did you think Meredith was pretty?" I asked. A frown briefly crossed Erin's face.

"Yeah, of course I think Meredith's pretty," she responded, as if any other answer were inconceivable.

I continued my line of questioning. "So does that mean you thought you were pretty?"

Erin's answer surprised me. "I definitely grew up thinking Meredith was prettier than me," she began, "because she's always been smaller than I am." Erin looks ashamed to admit this, as though she's somehow being disloyal to Meredith by saying it. She clarified that even though they may have looked the same to outsiders, it didn't feel that way to her. "In my family, there's still this idea that I'm the big twin. It's weird, but yeah, it's so funny, because we're identical twins, but I've always thought that Meredith is prettier than I am."

Erin spent much of her childhood feeling like an ugly duckling beside her beautiful identical twin. Because Meredith was smaller, Erin believed Meredith was prettier. Family members called Erin "the big twin," which Erin translated into "the fat twin" or "the ugly twin."

Erin started worrying about how she looked around age eight. In part, this was due to a burgeoning awareness that Meredith's appearance was more valued than Erin's because Meredith was thinner. But the twins faced another challenge. They hit puberty early—really early. By the time they were eight, they had developed breast buds and their parents required them to wear bras. Before Erin was even approaching junior high, the appearance of her body was transforming her life in noticeable ways, most of them unpleasant.

Erin and Meredith started getting "weird" looks from guys.

They didn't want that type of sexual attention, so they dressed like boys to downplay the puberty-related changes their bodies were going through. They wore their dad's football jerseys and big, loose T-shirts. They started getting mistaken for boys in public. All Erin and her twin wanted was to play football and soccer, get dirty, roughhouse, and do all the other things they used to do. But the way their bodies were developing complicated all of those activities. Erin talked about becoming more aware of her body when she was out playing with other kids. "'Cause it hurt," she explained. "If you would run, or things like that, your breasts hurt." But it wasn't just the pain of roughhousing while her breasts were growing that affected Erin; there was something more disturbing at work as well.

"When we'd play," she explained, "boys would land on you in a certain way to touch you."

"On purpose?" I asked, perhaps more shocked than I should have been.

"Yeah." Erin continued, a bit of anger boiling up as she recalled those moments. "On purpose. I was really, really hyperaware of how my body would interact while I played. It's hard to play when there's a part of your brain that's monitoring, like, 'Oh, did I show my boobs when I bent over?' And I'm, like, nine years old. It was weird."

Erin made it clear that when she and her twin dressed in boys' clothes, it had nothing to do with the development of any sort of transgender identity. She and Meredith both identified as girls then, and they identify as women today. Hiding their bodies just seemed like the only way to keep living the lives they wanted to live. By age nine, Erin was already wrestling with a sense of shame over the shape of her body. She was experiencing a growing recognition that if she didn't hide her body, it would be seen as sexual by other people and she would have little control over how they responded to it.

What Erin was wrestling with is nothing new. In an article from the 1970s about how women are portrayed in art and film, a British film theorist named Laura Mulvey argued that the essence of woman could be described as "to-be-looked-at-ness."[1] The term doesn't exactly roll off the tongue, but it does capture something extremely important. Erin felt "on display" as she walked down the streets in her neighborhood, and when she was at school, and when she was out playing with her sister. Before she was even ten years old, Erin understood that walking around in a female body carried with it a degree of "to-be-looked-at-ness." But just like the client I mentioned at the opening of this chapter, Erin didn't want to be looked at. That feeling she had of being on display started limiting the ways she could live her life.

A few years ago, I conducted an online survey of several hundred eighteen- and nineteen-year-old women around the United States about body image and related issues. Over 40 percent of these young women said they "frequently" or "almost always" felt like someone was evaluating their appearance. They felt on display, just like Erin. That feeling isn't a sign of neurosis or paranoia; it just signals that a girl or woman is paying attention to the world around her and being affected by the lessons she learns about her body.

The Shared Social Experience of Being Treated Like a Body

In 1997, two researchers, Barbara Fredrickson, who was then at the University of Michigan, and Tomi-Ann Roberts from Colorado College, published a theory that finally provided a big-picture perspective on the way our culture's focus on girls' and women's appearance is making them sick.[2] They called it *objectification theory*. It's the framework that guides my own research on beauty sickness.

Objectification boils down to this: It's what happens when you're not treated as an actual human being with thoughts, feelings, goals, and desires. Instead, you're treated as a body or just a collection of body parts. Even worse, your body is seen as something that exists just to make other people happy.

You lose your subjectivity when someone treats you like a thing, when you're seen as useful only if you can please someone with your appearance. Think of subjectivity as your internal reality. Your sense of self. The values and ideas and qualities that exist independent of anything going on around you. That's what's at stake here.

Objectification is inescapable for most women. It becomes routine and its results are insidious. When the focus is on the shape of your legs, no one cares about your intellect or your ambitions. When the world is deciding whether your body weight is acceptable, it's not worried about the kind of person you want to be or the kind of work you want to do.

Objectification can explain *why* the beauty ideal for women packs such a punch. Why is it that Artemis, Erin, and Gabrielle—women who grew up and live in very different circumstances—all suffer in similar ways? The driving force behind their experiences is objectification, that "to-be-looked-at-ness." It is only because girls and women know that their looks will be continuously scrutinized that the beauty ideal has the power it does. Otherwise, falling short of that ideal wouldn't feel like such a big deal.

Having the experience of being treated like nothing more than your body is unavoidable for most women. It's a shared social experience all women know. No matter how different we are, no matter what our backgrounds, we know what it's like to be objectified.

Street Harassment as Objectification

Once, when I was teaching, I used the term *leering* to describe objectification. In response, a student nodded and said, "Right, elevator eyes." She demonstrated on a giggling classmate, with a gaze that started on the "top floor" of the body, slowly crawled downward, then headed up again to rest on the breasts. Women in the class nodded. They knew this look.

The technical term researchers use for this kind of looking is the *male gaze*. It's fair to ask why the phrase specifically refers to men. After all, women look at men (and other women) this way sometimes. Nonetheless, we know from years of research that men do more of this type of looking than women do, and that women are much more likely to be the targets of this type of gaze. This visual inspection becomes even more disruptive when it's accompanied by commentary, ranging from the relatively innocuous "Lookin' good!" to explicitly sexual language.

Researchers at the Ohio State University developed an assessment of how often women are on the receiving end of this type of sexually objectifying behavior.[3] One of the questions asks, "How often have you heard a rude, sexual remark made about your body?" When I included that question in the online survey of young women I mentioned earlier, over 70 percent indicated that they had been the target of those kinds of remarks. Erin would count herself among that group.

A New York girl, Erin still lives in the same Bed-Stuy neighborhood in which she grew up. Erin was terrified when she was first catcalled at age twelve, while she was walking to the store with her sister. A stranger yelled out the window of a passing car. She doesn't remember what he said, just the fear and confusion she felt. No longer hiding under their dad's jerseys, Erin and Meredith looked years older than they were.

After that first catcall, Erin became highly sensitive to how men on the street looked at her or commented when she walked by. To make it even worse, Erin and her twin sister were growing fast. As a result, their clothes were often tight, which drew even more attention—though Erin notes that two girls walking down Fulton Street could just as easily get catcalled if they wore baggy pajamas. Being young and female was enough; no special attire was required.

Street harassment, the term that has emerged for these types of incidents, is a primary form of interpersonal objectification. Interpersonal objectification comes in many varieties, but street harassment is worth exploring in more detail because it provides a microcosm of the myriad objectifying experiences women face in their daily lives.

Street harassment has gotten a lot of attention in recent years, in part because of a 2014 viral video of a woman (actress Shoshana Roberts) walking around the streets of New York. As Shoshana walked, she experienced a shocking variety of catcalls and harassment. The video was conceived of by Hollaback!, an organization that fights street harassment. It sparked a national conversation about the topic that still continues today.

In a survey of almost five thousand U.S. women conducted jointly by Cornell University and Hollaback!, 85 percent of women reported first experiencing street harassment before the age of seventeen. Results were similar for a follow-up study that included women from an additional twenty-one countries around the world. When *Huffington Post* editor Alanna Vagianos recently called for women's stories of the first time they experienced this type of harassment (#FirstTimeIWasCatcalled), hundreds of women responded by sharing upsetting tales of their own experiences of being treated like sex objects by strangers. Many, like Erin, were as young as twelve.

The most interesting reactions to that viral video of street harassment were from those who insisted that most catcalls are meant in kindness. They're about appreciation, not harassment or intimidation! Women should be grateful for this attention, they claim. When the four women who host the Fox News program *Outnumbered* decided to argue in favor of the supposed psychological benefits of street harassment, Arthur Aidala, guest panelist and well-known criminal defense lawyer, proudly demonstrated his "signature move" for street harassment. He stood up and gave a slow clap, while reenacting the way he leers at attractive women he sees on the street.

Aidala clearly saw nothing problematic about his behavior, viewing it as a sweet compliment at best, harmless at worst. Jessica Williams, a correspondent on *The Daily Show* at the time, did a fantastic takedown of Aidala's arguments. After explaining that the sidewalk isn't a fashion runway or a red carpet, she noted, "Since going to work isn't a performance, we're not looking for applause."

Williams was spot-on with her use of the word *performance*. When objectification becomes a routine fact of life, girls and women do learn that to some extent, anything they do in public may feel like a performance. They learn to preemptively smile so that strangers won't demand they do so. They struggle to find that slippery boundary between being attractive enough to be accepted, but not so attractive as to draw dangerous and unwanted attention.

Erin struggled with that boundary as she moved into tweenhood. There were lots of conversations between Erin and her parents about clothes, many of which Erin found confusing. Her parents would tell her that her skirt was too short or that a top was too tight. Erin frowned as she recalled these conversations. She explained that at that age, she didn't even really understand what those comments meant. Too tight for what? Too short for what?

One time, Erin's friend gave her a stretchy shirt Erin really liked. Erin came down the stairs in the morning wearing the top, and her mom "freaked out." Erin explained, "My mom was really upset that I was wearing it and told me I looked like a slut, all of that. And I'll never forget, she actually came to school later that day, in the middle of the day, and pulled me out of class and apologized to me for shaming my body."

"What did she say? How did she explain it?" I asked, curious how a parent could even begin to explain the complexities of the situation to a preteen.

Erin remembers the moment clearly. "She said that it wasn't a problem of who I was when I wore that shirt, but it was a problem because if I exposed myself, I exposed myself to danger. She said, 'I don't really want to talk to you about this danger when you're twelve, I want you to be able to be a little girl, but if you wear this outfit, people are going to interact with you like you're a woman, and that's not always the best thing.'"

"Did that stick in your head, that there's something dangerous about being a woman?"

"Yeah," Erin agreed. "Absolutely."

Take a moment to think about the lessons Erin likely picked up from this interaction. She learned that looking like a woman, that *being* a woman, was a safety issue. She learned that her fashion choices could put her in danger and thus should be monitored closely. She learned that she needed to spend *more* time worrying about how she looked to other people—a terrible lesson, but one that most parents can't help passing on to their daughters. Like any other girl, Erin wanted to be pretty. But she learned that if pretty went too far, it could become sexy, and being sexy wasn't safe.

When I asked Erin how she thinks her parents handled these interactions, she had nothing but sympathy for what they were going through. She noted that parents back then didn't have the resources they have now, and she's right about that. There was no A Mighty Girl to provide encouragement to parents seeking to raise healthy, body-positive daughters. The landmark American Psychological Association report on the sexualization of girls was still more than ten years away. Even the most thoughtful and careful parents were flying blind to a large extent.

How do you teach your daughter that there's nothing wrong with her developing body, but tell her she still needs to hide it sometimes if she wants to be safe? How do you teach your daughter positive, healthy attitudes about sexuality if the first things she learns about her own reproductively maturing body are that it both puts her in danger and simultaneously gives her the power to entice grown men? These are dangerous waters, difficult to navigate. It's hard to come out on the other side with a healthy respect and appreciation for your body that isn't tied to men's appreciation of it.

While the particulars of Erin's story might be unique to her, her more general experiences are not. Sofia*, a nineteen-year-old Mexican American, saw my call for interviewees and agreed to meet with me when she was in Illinois visiting friends over her summer break. Sofia seemed shy when she first arrived at my office, shifting in her chair and pulling at the cuffs of her shorts. Her voice started out so low that I struggled to hear it over the din of construction going on outside my office window. But her volume increased when we started talking about catcalls. She sat up straighter and her gaze became more intense. Sofia was angry, and she had some things to say about street harassment.

"Because it's, like, literally, there's nothing I can do about it," So-fia began. "Is there anything I can do about it?"

Sofia wasn't really asking me for an answer, she seemed to be aiming the question out at the universe in general. Nonetheless, I followed up. "What do you think?" I asked.

As I suspected, Sofia seemed to have already made up her mind. "You know," she told me, "it's always gonna happen."

"Always?"

"Unfortunately, I think it is," she confirmed, with a note of finality.

Though Sofia's normal strategy is to ignore catcalls, she told me of one incident when she decided to speak up. Sofia's a student at an East Coast college, and she was on her way to a formal dance with her boyfriend. She was wearing high heels and a pretty skimpy dress, definitely feeling sexy. She'd been excited to wear the dress for months. When she walked past a group of construction workers, one of them shouted, "Look at that big ass!"

Sofia's more confident about her body than your average young woman, but this guy was hitting where it hurt. She worried about the size of her butt, so that comment didn't feel like a compliment, even if it was intended as praise. The joy of wearing her new dress was driven away by anger, and she began feeling self-conscious instead of confident. Because the construction worker was on a roof, Sofia felt safe speaking up. She knew he couldn't get down from the roof fast enough to hurt her, so she told him to "go fuck himself." It felt good to say that, Sofia told me. But it didn't bring back that breezy feeling she'd had before this stranger decided to comment on the shape of her ass. And it didn't change the fact that for the rest of the evening, she couldn't stop worrying about how her butt looked in that dress. That one comment reverberated in her mind for hours afterward.

I was surprised to hear the incident happened when Sofia was walking with her boyfriend. Sofia laughed. "Yeah," she confirmed, "I was with my boyfriend, and he almost didn't know how to react. I guess he had never seen a girl get catcalled before."

"Were you, like, 'Welcome to my every day'?" I asked her, laughing along.

"Yeah," Sofia sighed. "Every day."

That story has a relatively happy ending compared to the tale Sofia told me about the ice cream bar. It started out simple enough. Pretty late at night, Sofia was watching the movie *Barbarella*, which she described as "this really sixties movie with Jane Fonda as a bride who's in space."

Sofia wanted an ice cream bar. But she also saw how good Jane Fonda looked in the movie. So she struggled. Could she have an ice cream bar? Did she *deserve* an ice cream bar? Finally, Sofia made her move, leaving her apartment and walking down the street to a 7-Eleven. In the store, a customer by the freezer section started catcalling her. He told her she was sexy. He detailed what he wanted to do to her. Sofia grabbed her ice cream and headed to the counter to check out, but the man followed her and stood in line close behind her. She wanted to leave, but she needed to pay. Sofia remembers thinking that she was not going to let some asshole deprive her of the ice cream sandwich she had already spent so much time thinking about. She followed her general harassment protocol, which is to pretend the guy doesn't exist. She didn't talk to him or look at him. She didn't want to give him the satisfaction of seeing a reaction out of her.

Sofia still remembers that the ice cream cost $2.33. The cashier saw what was happening and tried to quickly herd Sofia out of the store. He took her two dollars but told her not to worry about the thirty-three cents. Sofia left. The man didn't follow. Sofia got back home safely and ate her ice cream. But afterward she felt guilty for eating that ice cream bar. It somehow seemed that what happened in the 7-Eleven might have been punishment for her for splurging. "It was a bittersweet ice cream," she recalls.

I could tell when Sofia shared this story that she still wasn't sure how she was supposed to have handled the situation. She also wasn't sure if that cashier did her a favor or if he was guilty of not intervening sooner or more forcefully. She shifted in her chair in my office. Shook her head. Shrugged her shoulders as if to say, "What are you going to do?"

Though the risk of sexual assault is all too real, street harassment and catcalls aren't important just because they make women feel vulnerable to assault. They also serve as ongoing reminders to women that their bodies are on display every time they leave the house. As Erin explained it to me, you pay a price just for being a woman who doesn't want to be completely invisible. "Scrutiny," Erin described the feeling. "Constant scrutiny. When you leave your house, you are under constant scrutiny. And you can either crumble from that pressure—which a lot of women do, of course, and you can't blame them for it. Or you can just totally comply with it. Or you can be like me and be pissed off all the time. Or you can try not to care." These don't seem like great options for women when it comes to dealing with the scrutiny that breeds beauty sickness.

At twenty-two, Erin took a radical step that ended up having an unanticipated effect on her experiences of street harassment. At that time, Erin had just broken up with a boyfriend who had been treating her badly and controlling every aspect of her life. She felt drastic measures were needed to regain the sense of being in charge of her own destiny. She asked herself, "How am I going to gain control back?" The desire to regain that feeling of control was intimately tied to Erin's feeling that others behaved as though they owned her body. Erin explained that feeling to me. "When you're a woman, it's like your body is everything: your body is the first thing people notice and talk about; it's the first thing people gossip about; it's the first thing people point out about women."

At the time, Erin's hair was "long and very beautiful" and she had grown it out that way for this boyfriend. After they broke up, Erin shaved it all off as a way to reclaim herself. Went completely bald. She started wearing oversize T-shirts every day, just as she and her sister had during that difficult period of childhood. Erin also let go of her makeup, but first constructed a ritual around it. I was fascinated by this part of her story.

"You threw away all your makeup?" I asked, somewhat incredulous.

"I did," Erin confirmed. "I wrote letters to them and threw them away."

"You wrote letters to your makeup?" I smiled, loving that Erin showed such respect for the symbolism of that moment.

Erin painted the scene for me. "Yeah, so I told them . . . I wrote, 'Eye shadow, you're bad for me. You make my eyes sting and tear up and I feel compelled to wear you anyway, but now you gotta go. Good-bye, eye shadow.'" Erin bows her head and gently touches the table in my office, as though bestowing a benediction on that eye shadow. She continues, "And I wrote letters to all of them, and I wrote, 'You can be a guest in my home for one night, but tomorrow you're out,' and I kept them on the table overnight and then I threw them all away in the morning."

So there Erin was, bald head and makeup-free skin. She also committed to avoiding mirrors for a period of time. These choices changed Erin's life in some unexpected ways. She noticed a drastic reduction in street harassment. Erasing her more traditionally feminine appearance allowed her to step outside of the game. Men no longer saw her as a sexual object. To Erin, her baldness was a deliberate, powerful provocation. She had simply had enough.

Erin describes that time period. "The years I had a buzz cut or a Mohawk, it was definitely like a hostile performance, absolutely, absolutely." Erin nods vigorously, reliving that sense of power.

Not everyone appreciated Erin's decision. She remembers that every time she went somewhere, people would tell her, "You would be so pretty if you grew out your hair." But what was more interesting was the effect shaving her head had on her experiences of street harassment.

"What's really bizarre," Erin explained, "is that my street harassment became zero, none—no street harassment. When I had a shaved head, I got so much respect from men, it was amazing. Because they didn't see me as sexual at all. I wasn't even on their sexual radar. I was all of a sudden treated like a human being."

Erin identified the magic border between street harassment and no street harassment as somewhere between two and three inches of hair. That's all it took. It's been a few years since she shaved her head and said good-bye to her cosmetics. Erin has recently invited some makeup back into her home. She's also growing her hair out. I asked her about that decision.

"I'm growing it out now because I don't feel like I need to be so defensive," she told me.

"Is it worth it to grow it back?" I wondered aloud.

Erin was thoughtful. "I'm still asking myself that, for sure."

It's a strange world we live in, when a couple of inches of hair growth can fundamentally alter a woman's experience of moving around in the world. But it's consistent with everything we know about objectification. The more easily identifiable you are as a woman, the more you will be objectified. The more you are objectified, the more your body will begin to feel like a performance piece instead of the home in which you live.

Is Objectification Always a Problem?

At this point you might be thinking, "So what if men look at women and occasionally vocalize their pleasure?" Of course heterosexual men are going to be drawn to attractive women. Why wouldn't their eyes and their words reflect that attraction? Or you might find yourself thinking that sometimes it feels good to have someone appreciate how your body looks. Is there really anything wrong with that?

There are two ideas here worth exploring. The first is that many women, at least in some circumstances, feel flattered by the types of attention we might call objectifying. There are moments when we want our bodies to be looked at and appreciated. The second is the oft-cited notion that objectification may be common, but it doesn't have any meaningful impact on women's lives. The latter argument is easy to tackle. The negative consequences of the objectification of women are well documented. Objectification feeds beauty sickness in countless ways. But I'll address those outcomes in the next few chapters. For now let's consider the first idea, which is more complex.

How do we reconcile the outrage many women feel about being catcalled with the brief burst of self-esteem that *sometimes* accompanies compliments about one's body from a stranger? What makes some body commentary okay and other commentary threatening, or at least uncomfortable?

Curious about the answers to those questions, my lab conducted a study in which we asked women how they would feel in different objectifying scenarios. We varied the context in which the objectification occurred. In one scenario, it might be your romantic partner staring at your body parts. In another, it was a stranger doing the same thing. In some cases, we'd describe a scenario in which a stranger makes a sexual comment about your body, but varied whether it was happening when you were walking with friends or walking alone. Not surprisingly, women reported particularly negative emotional reactions to being objectified by strangers, especially if they were alone when it happened. That makes sense, given safety concerns. However, even though these young women said being catcalled by strangers was a generally negative experience, some admitted a few positive feelings can come along for the ride.

Sofia's experiences are consistent with these research findings. She recognizes both positive and negative outcomes to getting catcalled. It can make you feel pretty, but it can also be frightening, particularly when the comments are focused on sexual acts or specific parts of your body. Sofia described sometimes feeling "like a part and not a person" as a result of catcalls. Other times, a "Looking good, honey!" makes her day. She mentally responds, "Hell yeah, I'm looking good!"

Satirist Melanie LaForce captured the complexity of these feelings in a recent piece published in *Thought Catalog*.[4] She wrote, "Anyone who has felt unattractive at a given point can appreciate feeling attractive. Granted, there is a line between feeling that a man views you as attractive, and feeling that a man views you as a scantily-dressed hamburger, but it's a goddamn continuum. Some of us enjoy those hamburger moments, and YOU CAN'T TAKE AWAY MY HAMBURGER." The hamburger metaphor is perfect in some ways. A hamburger can be wonderful, but too many hamburgers over time can have a cumulative effect that threatens your health. And some hamburgers are delicious while others make you sick. But nobody wants a hamburger when they ordered something else, or didn't order anything at all. Likewise, no woman wants to be treated like a piece of meat when what she's really looking for is respect, or understanding, or even simply to be left alone.

But there are scenarios in which objectification might be welcomed by some women. Think of a group of women getting dressed up to head to a party with friends or for a night out at the bars. That social ritual often involves deliberately creating an appearance that invites others to look at you in a sexual way. Back when I was in my twenties, my girlfriends and I would talk about "going-out clothes." It wasn't some secret code. We all knew what going-out clothes were. They were shorter than what we would usually wear. Or tighter. Or lower cut. They were clothes designed to draw in the eyes of others. It can feel sexy to be looked at in that way. It can feel powerful. It can be fun. But remember, objectification isn't black and white. It exists on a continuum. Just because you're okay with one spot on that continuum doesn't mean there's not a point at which you'd begin to feel uncomfortable or afraid. More important, even if you are comfortable with some types of objectification in some situations, that doesn't mean it can't still have a negative impact on you. Choice is powerful. But just because we can choose to be seen as sexual objects at times, that doesn't mean we can opt out of the consequences that accumulate as we have more and more of those experiences. And making that choice on occasion does nothing to protect us from all of the times we're given no choice.

Putting too much stock in your own attractiveness can also leave you incredibly vulnerable to negative evaluations of your appearance. We have to be honest about the fact that knowing your physical appearance is always up for evaluation can make you feel you are being held hostage by others' opinions about how you look. There's a reason that being called ugly hurts women so much more than other types of insults.

Erin remembers being harassed on the subway once. The man was in her face, talking about how sexy she was. Erin told him she wasn't interested. She told him to "leave her the fuck alone." But she also refused to move to a different car. She didn't want to have to give up her seat. For the rest of the ride home, she had to listen to "You're an ugly ass bitch. You're so fucking ugly. I can't believe I was talking to such an ugly ass bitch." The irony of this situation isn't subtle. A man approaches you because he finds you sexy. Then when you reject his advances, he calls you ugly.

If we train young women to believe their most important asset is their appearance, of course men (and women) know where to hit when they want to inflict the most psychological damage. Erin says it best: "That's where you get it, because that's what matters the most. 'You're ugly.' That's the biggest knife you can wield around women." Erin's take on the matter is identical to seven-year-old Leigh's. Remember, even Leigh already knew that the worst thing you could say to a woman is that she is fat and ugly. It is no coincidence that when women say things men don't like, the response is often not a critique of their ideas, but a critique of their appearance. This is the logical outcome of seeing women as objects instead of as human beings.

What Happens If You're Always "The Hamburger"

It's unrealistic to think we could live in a world where looks don't matter. There are moments when objectification may feel pleasurable, and we may feel powerful when we can use our bodies to attract attention and admiration. But the feeling of being on display all the time does something to the psyches of girls and women. There are costs to so often being "the hamburger," even if we like the idea of being the hamburger once in a while. When other people are constantly focused on your appearance, you're bound to do the same.

Imagine you're a nefarious woman-hating villain and you're trying to develop a strategy to keep young women from reaching their full potential. But you need the strategy to be something subtle, a strategy that makes it appear as though women are freely choosing their own undoing.

Simply follow these two steps. First, evaluate women's bodies with such unremitting intensity that women feel they're on display at all times. Second, sit back and wait as young women internalize this perspective and learn to think of themselves as things to be looked at instead of as complete human beings with goals and dreams that have nothing to do with the size of their waist or shape of their breasts. The result of these two simple steps is the voice in the back of so many women's minds that's too busy asking, "Do I *look* okay?" to ask more important questions. That voice is what beauty sickness sounds like. The more you hear that voice, the more your body begins to feel like something that exists for other people instead of feeling like your home.

Of course, there's no archvillain running the show here. Things are not so black and white, and this isn't a zero-sum game. Plenty of women are busy asking plenty of important questions. But if you look around, it does seem that steps one and two above have been implemented with a good deal of success.

This is the crux of living in a world that so easily bypasses women's humanity to focus on their looks. Erin's battles with beauty sickness evolved in a dramatically different way from Artemis's and Gabrielle's. Nonetheless, all of these women are soldiers in the same war. They're all trying to find a way to live in their bodies without hating their bodies, to be a part of culture without being controlled by it, and to have the freedom to step away from the mirror and spend more time thinking about being something other than thin and pretty.

TWO

This Is What Beauty Sickness Does to Women

Your Mind on Your Body and Your Body on Your Mind

I MET WITH Ana* in a suburban Starbucks, the summer after she had just completed her first year of high school. Her dad dropped her off and signed a form giving me permission to interview Ana, then left us alone at our small table amid a sea of men in business suits. Ana identifies as a multiracial Latina, but is generally assumed to be white by those who meet her. She is on the short side, with a small frame made more imposing by an unusual amount of poise and maturity for her age. Ana recently dyed her hair bright pink. Clearly pleased with the result, she describes it as giving off an "intimidating vibe" that makes her bolder.

Ana didn't know me at all when we met at that Starbucks. The beginning of a new school year was right around the corner, so I tried to start our conversation by asking Ana if she had thought about what she would wear on the first day of school. She frowned at me. I got the sense that if she had been less polite, she might have rolled her eyes as she said, "Um, I don't really plan those kinds of things in advance, you know?"

Ana explained that her style shifts with her moods, making it a waste of time to plan ahead. But she does spend a good deal of time thinking about her style and how she presents herself to the world. Ana finds fashion compelling and loves playing with cosmetics. She regularly watches YouTube tutorials on new makeup techniques and practices on herself and her friends. The day I met her, her eye makeup was as bold as her hair. It gave her a sharp, skeptical look.

Fifteen-year-old Ana self-identifies as a feminist. She's intellectually precocious, casually tossing around words like "oppression." She also has an unusually healthy attitude toward her body for a young woman her age, explaining, "There's this idea of a perfect body out there, but that wouldn't necessarily be the body I would want. I think saying that you wanna choose to have that, is just, like, so unaccepting of your body."

I've heard other women say body positive things like that, but I'm not always convinced. Often when I hear those sentiments, they sound forced—more about how a woman wishes she felt than how she actually does feel. But Ana seems to mean it. I believe her. And I'm happy for her. It's refreshing to talk to a young woman who seems pretty comfortable in her own skin.

Despite her confidence, Ana still occasionally finds herself on a beauty tightrope. She admits that she worries about what other people think about how she looks, even though she wishes she didn't. The other day Ana caught herself thinking she shouldn't leave the house without doing her eyebrows.

"Would people think differently about you if you didn't do your eyebrows?" I asked. "What do you think?"

Ana cracked up and responded with a dramatic flair, "Well, from a technical standpoint, eyebrows *are* what frame your face!"

I'm glad she can find humor in the absurdity. If you're going to be bound by rigid beauty ideals, meeting them with laughter is probably preferable to meeting them with tears.

Even though Ana loves fashion and has fun turning her hair a rainbow of different colors, she worries about letting her looks define her. She doesn't want to be "that girl." Ana recognizes a gray area between beauty practices designed to express your style and personality and those you feel you have to do in order to appear acceptable to others. She talked about the difficulty of "figuring out if you wanna wear makeup 'cause *you* wanna wear makeup, or you wear makeup because you feel like you *have* to."

Although Ana wants to believe that all her beauty behaviors are freely chosen means of self-expression, she wonders if it's even possible to tell the difference between when you're trying to look good for yourself instead of doing so for others. How do you know that your idea of what it means to look good is really your own, since that idea could never develop without input from your beauty-sick culture? Maybe Ana's pink hair is a statement of individuality and confidence. Maybe it's a big middle finger to mainstream beauty ideals. But Ana recognizes that it could also be something that, in the end, draws more attention to how she looks and more commentary about her appearance from others.

Ana doesn't want to live in a world where people view her as more about how she looks than what she's saying, but she fears that may be the case. "It's the expectation that women are, like, objects. And if they're not pretty and beautiful, then what are they worth? It's a really bad expectation, but it's still an expectation that women look good constantly." Then Ana hits the nail on the head. She tilts her head to the side, sets down her drink, and looks me in the eyes. "I'm concerned a lot. I worry, what if *I* view me as more about my looks than what I'm saying?"

Becoming Your Own Surveyor

What Ana is describing is something called self-objectification. Girls and young women learn quickly that their appearance is going to require and attract near-constant attention. Over time, the awareness that others are evaluating your looks (or could be at any moment) gets internalized. If other people are always monitoring your appearance, eventually you'll start doing the work for them. You become the closest observer of how you look, a constant surveyor of your own body. For this reason, self-objectification is also referred to as *body surveillance* or *body monitoring*.

Think of self-objectification as something that carves off a little piece of your awareness so that you can monitor how you look to other people. *Am I sucking my stomach in? Are my pants riding up? Is that pimple on my chin showing? Does my hair look okay?* There's nothing weird or unhealthy about having an internal monologue. It's an important feature of human consciousness. But try to put it in context by imagining other types of questions you could ask yourself as part of an internal monologue. *Am I making the best decision? What am I capable of learning today? How am I feeling? What do I need? What do the people around me need?*

Remember, we've taken an important step here; we've gone from a culture that reminds you that your body is being looked at to *your* being the most consistent surveyor of your own body. When I explain body surveillance to women, I usually talk about what it's like to look in a full-length mirror. Imagine you're doing that now. You're taking in how your clothes fit, what your skin looks like, how well your hair is conforming to your wishes. Self-objectification is what happens when, even though you're no longer in front of that mirror, you might as well be. It remains in your mind. Because as you're walking around, sitting at a table, or listening during a meeting, a part of your consciousness is still monitoring how you look. When you're giving a presentation in front of a group, a part of you is sitting in the front row, checking yourself out, deciding if you look okay. Never mind what you're actually saying. Is your skirt too tight? Is there lipstick on your teeth?

Style.com (perhaps inadvertently) captured the idea of chronic self-objectification in a campaign called "Runways Are Everywhere." The campaign's images showed female models strutting their stuff in a variety of everyday locations. One widely critiqued image shows a model wearing nothing but a bikini, boldly posing inside a Laundromat, next to a row of washing machines. Think about the message being communicated with that image. As a woman, even if you're just hanging out in the Laundromat, you should assume your appearance is on display. Every moment is a beauty pageant.

To some extent, self-objectification can act like a personality trait, remaining relatively stable across situations. Some people are just more prone to self-objectification than others. No matter what the context, no matter who they're with or how they're feeling, they're monitoring their appearance. Researchers out of the University of Wisconsin created a measure of this type of self-objectification with items like "During the day, I think about how I look many times" and "I often worry about whether the clothes I am wearing make me look good."[1] You probably know women who would strongly agree with those statements; you might even be one of them. But even someone who isn't prone to these everyday high levels of body monitoring can still be pushed into self-objectification by the right context. A glance at a magazine cover, a comment from a colleague. It doesn't take much to make that mental mirror materialize.

I still remember one of the first sets of teaching evaluations I received, back when I was a twenty-four-year-old graduate student instructor. I loved teaching and was ridiculously nervous about reading my students' feedback. I wanted my students to like me and my class as much as I liked teaching them. I've long since forgotten most of the comments in that set of evaluations, but there is one I can't seem to forget. It read, "You should wear that blue skirt more. It looks good." I was floored by that comment. The teaching evaluations were anonymous, but I couldn't stop mentally thumbing through the Rolodex of students in that class, trying to figure out who had written about my skirt instead of about my teaching. Who was thinking about my legs when I was thinking about how to best teach psychology? What's worse is that the class was "The Psychology of Gender." I felt I must have done something wrong as an instructor if, after an entire semester tackling the nuances of gender, one of my students still felt that kind of comment was acceptable.

That comment still bothers me today, even though I've waded through thousands of student evaluations in the years since then. I can picture the skirt in question. It was dark blue corduroy. I wore it with black tights and boots. That one comment had a lingering effect on my teaching throughout the next semester, perhaps even beyond that. I felt "looked at" in a way that I hadn't before I read it. I was unsettled. When I talked to men in my department about the comment, they blew it off. "It's a compliment," they said. "What's the problem?" But it didn't feel like a compliment. In my mind, that student might as well have written, "I respect you so little as a professor that I won't even address your teaching; I'll just make a reference to your wardrobe and body shape."

I didn't want my students to think about my body while I was teaching, but that wasn't the most serious problem with that comment. What was more serious was that *I* didn't want to think about my body when teaching. I have only so many attentional resources to go around when I'm in front of a class. I'm not interested in dedicating any of those resources to monitoring how my body looks. The problem with having your body on your mind is that it takes your mind away from other things.

In "Shrinking Women," an award-winning entry for a poetry slam competition, college student Lily Myers spoke of her own chronic body monitoring. In her compelling spoken word account of how she learned that being a woman meant learning to focus on her appearance, she shared, "I don't know the requirements for the sociology major because I spent the entire meeting deciding whether or not I could have another piece of pizza."

Beauty Sickness Eats Up Your Mind

I got in contact with Rebecca* through a former student who recommended her for an interview. A twenty-seven-year-old nurse, Rebecca agreed to meet at a hipsterish little café around the corner from the Baltimore hospital where she's employed. Rebecca works mostly with psychiatric patients, so it was a bit of a reversal for her, being on the telling side of a story instead of the listening side. But she quickly warmed up, and we monopolized our table in that little café for almost two hours as Rebecca told me her story.

Rebecca remembers a childhood relatively free of beauty sickness. I asked her, "Do you feel like you liked how you looked when you were a kid?"

Rebecca paused thoughtfully, seeming to scan back through her memories. "I don't think I thought about it. I don't remember thinking about how I actually looked, or my face or my overall appearance. I didn't think about what I wore."

"What did you wear?"

Rebecca grinned. "Whatever my mom gave me," she explained. "Stretchy pants and a Lion King sweatshirt?"

Rebecca's grin was so contagious, I can feel myself smiling while I type this, remembering that moment. I picture young Rebecca in a Lion King sweatshirt down to her knees, carefree and happy. But Rebecca's smile didn't last long. That period of childhood when Rebecca was free from appearance worries seems so short now, compared to the years she has spent hurting. By the time she was in junior high, Rebecca found herself thinking a lot about how her looks compared to other girls'.

"What do you think changed?" I asked her. "What made you start thinking about how you looked?"

It all started with Rebecca's teeth, which grew in at odd angles and begged for braces. Rebecca recalls, "So I went to the orthodontist, and he talked about changing my smile so I could have a movie-star smile. I remember that drawing attention to the fact that I could look a certain way. He asked, 'Whose smile do you admire?' And I remember there was a girl in school who I thought was pretty, and I said, 'I would like to have a smile like hers.'"

Rebecca was a bit of a late bloomer. She noticed that she was the last one to get a bra, the last one to shave her legs, the last one who wasn't "doing her hair." She had a sense that she was losing a race, even though it wasn't a race she was particularly interested in running at the time. She describes those days: "I thought, 'Everyone else is doing this thing, I'm behind, I gotta do it too.'"

Though Rebecca was never interested in makeup, she noticed that her friends were wearing it and thought, "Oh, well, other people are wearing it, I guess that's what you're supposed to do." When other girls seemed to care more about their appearance than Rebecca did, Rebecca simply assumed she should care more.

But the real turning point for Rebecca was at the beginning of high school. She started to hear comments about her appearance from others, which directed her own thoughts to her body. Sometimes the comments were subtle, overheard. Other times they were oddly official and openly shared, as when Rebecca's boyfriend at the time told her about something his teammates had done. She explained, "They had gone through and created a perfect female by taking segments of each girl that they thought won a category. So this person's face, this person's chest. And then he told me I had won the torso, which was the abdomen and butt."

I couldn't hide my distaste. "What did you think when he told you that?" I asked.

Rebecca is strikingly honest about her mixed feelings. "I was flattered and also a little bit disgusted by the idea. And the coach, who was a young guy himself, talked about who would be the best-looking in the future and they had picked me. Like this person would become most improved or something over time."

"What did it feel like to know that there was this group of guys who were talking about how you looked?"

Rebecca thought that perhaps that kind of behavior was normal. She told herself, "Oh, this is what the guys talk about." But she also explained, "I guess that's where maybe some of my own thoughts about myself started. My own face, my own body—like, I started to care."

What Rebecca is describing is a classic pattern. The more you feel others are looking at your body, the more aware you become of how you look to other people. The more the people around you seem to care about appearance, the more you feel you should do the same. Rebecca's body became a collection of parts for others to evaluate. This led her to spend more time doing that same type of evaluating herself. Objectification turned into self-objectification.

Though we like to imagine ourselves as masters of multitasking, the truth is, we are not. If your attention shifts toward your appearance, it's shifting away from something else. Rebecca knows what this is like, because throughout her adolescence and well into her twenties, a significant proportion of her mental energy was allocated to tracking her food and exercise habits with gut-wrenching precision. It all started during her high school health class, freshman year.

Rebecca was a competitive swimmer, fast and powerful. Her body was tall and muscular. The trouble began when, like many students across the country, Rebecca had the experience of having her body fat measured as a part of health class. Her health teacher told Rebecca that she measured "obese." In that same class, Rebecca was taught that 2,000 calories a day was the "right" amount to eat. No one told Rebecca that a competitive athlete—especially a swimmer—needed many more calories. And no one told her that those types of body fat tests are notoriously inaccurate, especially for athletes. Certainly no one taught her to question the validity of the categories commonly used to describe different BMIs.

So Rebecca began to worry. Things just didn't seem to add up right when it came to her body. She worked on the math more and more. Adding exercise, subtracting food, counting calories, carrying around a journal where she charted eating goals. Her mind turned into a spreadsheet with running totals updated as she moved throughout her days.

When I asked Rebecca to give me an estimate of the mental space this spreadsheet took up, the question itself overwhelmed her. Even as she looked back, it was a lot to get her brain around. To give me some context on how immersed she was in her mental spreadsheet, she described once noticing a woman who had normal eating habits.

"I was kinda fascinated by it," Rebecca began. "Like, 'Oh, she just eats when she's hungry, and if she's not hungry, she just won't have something. But it's not because she's not eating, it's just because she's not hungry?' I was curious about it. And jealous, when I realized she didn't really have to think about it."

"How do you think your life would've been different if you would've felt like that girl did?" I asked.

Rebecca tells me she would have had "just a lot more, like, mental space. Like, my mind would've been free for a lot of other things." It was hard for Rebecca to devote her mental energy to anything else because, as she explained, "I just had a constant calorie meter. Throughout the day, I'd be watching it go up and down. I just couldn't stop, like a bad habit that I couldn't shake."

Rebecca told me about a time when she was a teenager and went shopping with her mom, who was looking for a new bathing suit. Rebecca's mom had recently gained a few pounds and struggled to find a suit she liked. She casually remarked to Rebecca, "Ugh, this is the first time I can ever remember being unhappy with my body." You can see it in Rebecca's face that she's still blown away, just remembering this offhand comment her mother made. She recalls her thoughts at the time: "I'm, like, 'That's amazing. That you're forty? You're forty when you felt unhappy with your body for the first time? You made it that long?'" Rebecca still looks startled by this, incredulous.

It's hard not to compare Ana's experiences with Rebecca's. When I asked Ana how she feels about her body, she said, "Well, everyone has things they wanna change." But then she couldn't come up with anything she wanted to change, which was nice to hear. Ana says she has "a pretty good handle on things" when it comes to body image, but she still occasionally feels bad after seeing media images of idealized women's bodies.

"What do you do if you feel bad?" I asked Ana. "How do you get yourself out of it?"

Unlike Rebecca, whose response to body worries was to catalogue and monitor all of her areas of concern with extraordinary dedication, Ana says she simply "focuses on something else" when she feels bad.

"Seems like a good strategy," I say.

"Yeah." Anna confirms. "Honestly, I feel like dwelling on it makes it worse."

Ana's right. Dwelling on the appearance of one's body does make it worse. Body surveillance has unfortunate consequences for your brainpower. It is especially likely to interfere with cognitive performance when a task is difficult and requires you to focus. In other words, the times when you most need all of your focus and attention are when body monitoring has the potential to be most disruptive. Imagine the implications this has for girls and women in educational settings. How much of a hit are schoolwork and test scores and the ability to learn new, important, or exciting things taking when girls are distracted by their own appearance monitoring? As seventeen-year-old Artemis (from chapter 1) put it, "It's like my brain is split into two halves. Half is about my body and half is school."

Several researchers have induced self-objectification in a lab setting by asking women to try on a bathing suit privately, but in a room with a mirror. For most women, the very act of trying on a bathing suit was enough to draw their attention to the size and shape of their body. But even more important, one of these studies demonstrated that this simple act of trying on a bathing suit resulted in lower scores on a math test. The study was aptly titled "That Swimsuit Becomes You."[2] In a different bathing suit study titled "Body on My Mind,"[3] researchers found that trying on the suit unsurprisingly led to body shame. But even ten minutes later, after the women redressed, they still had lingering body thoughts. It's easy to get women to start thinking about how their bodies look. Once your brain heads down that road, it's hard to change direction. Remember, in these studies, no one actually saw these women in the bathing suit. But they saw themselves, and that was enough.

The Stroop test is a task researchers often use for studies focused on mental resources. It's straightforward and the instructions are easy, but doing it well requires concentration and focus. The task involves looking at a long list of words and quickly identifying (out loud) the color in which they are printed. It sounds pretty easy, and it is, unless the words are color words that don't match the color of the ink. For example, it's really easy to look at the word *dog* printed in purple ink and say "purple." It's a lot harder to see the word *yellow* printed in blue ink and say "blue." Your brain wants to say "yellow" instead. You read the word without even meaning to. Doing many of those mismatched trials is mentally exhausting. A study published in *Psychology of Women Quarterly* demonstrated that trying on a bathing suit actually causes women to perform worse on the Stroop test, presumably because a portion of their cognitive resources has been stripped away to be used for body monitoring.[4]

Women who are habitual body monitors seem to be the most sensitive to these types of disruptions in thinking and attention. This makes sense. Think of it this way: If you're a chronically happy person, it's pretty easy to get you to laugh at a joke. In the same way, if you tend toward chronic body monitoring, it's pretty easy to distract you from whatever else you're doing and get you thinking about how your body looks. Researchers from Yale found that women who spend a lot of their cognitive resources monitoring their body also report lower levels of motivation and lower self-efficacy.[5] If you want young women to grow up ready to make a difference in the world, self-efficacy is invaluable. It's the belief that you can do what you set out to do, that you can control your own behavior and have an impact on the world around you. We shouldn't have to sacrifice our valuable focus and attention in exchange for body monitoring. Our minds have too many more important places to be.

Clothing That Wears You Out

When I was around seven years old, I went to a party with my parents. It was a good time, with pizza, ice cream, and lots of young kids around to play with. I remember wearing a purple quilted dress my grandmother made for me, with a puffy white blouse underneath. When some other children started doing somersaults down the hall, I happily raced to join them. I can still hear the voice of the adult who pulled me aside and told me I couldn't play like that because I was wearing a dress and it wasn't ladylike—people might "see things." I felt confused and chastised. I sat quietly on a chair by myself, "like a lady," watching the other kids tumble by. I learned a lesson that day that so many little girls learn—to associate being ladylike with sitting quietly on the sidelines of life. I learned I had to pay attention to how others might perceive my body.

This type of disruptive body monitoring is often caused by (or at least reinforced by) a tendency for girls' and women's clothing to be more restrictive than boys' and men's clothing. A necktie may be uncomfortable, but it comes nowhere near high heels or a pencil skirt in terms of restricting how you can move your body.

I'm dating myself here, but one of the images that always comes to mind when I think about clothing triggering self-objectification is the famous green dress worn by Jennifer Lopez to the Grammys in 2000. I'm not the only one who remembers this dress—it has its own Wikipedia page and is on display in a fashion museum. The Versace gown boasted a wide and deep V-cut in the front, going well below the navel and barely covering Lopez's nipples. Reports after the Grammys said that J.Lo employed double-sided tape (perhaps in addition to some luck) to keep the dress in place during the awards ceremony. Here's what I keep wondering: Does J.Lo remember anything from that night? Could she focus on the speeches, the performances, the friends sitting around her? Honestly, how could she think about anything besides whether her breasts still had just the right amount of coverage? What happened when she sat down? Did she have to suck her stomach in? Can you even sit down in that dress? How much of a distraction must wearing that dress have been?

A few years ago, I was eating dinner in a little Italian restaurant near my home. A group of ten or so high schoolers came filing in, dressed to the nines. Based on the time of year, I assumed it was a group eating dinner before their annual homecoming dance. I could not help but watch the girls as they walked by and took their seats. They looked lovely—they did. But their dresses were incredibly short. So short that I wondered how exactly they were going to sit down without flashing everyone. The tables in this restaurant didn't have tablecloths, so there was no hiding once you sat down. These young women could barely eat, because they had to keep using their hands to tug down the hems of their dresses. Those wearing strapless dresses had double duty: tug down the bottom, yank up the top. Take a bite of food. Tug. Yank.

You don't have to be wearing a strapless dress or something cut down to your navel to find yourself in the position where keeping your clothing in place distracts you from what's going on around you. For girls, this experience starts shockingly early. After observing several preschool classrooms, a sociologist out of the University of Michigan noted that dresses clearly interfered with how young girls could move.[6] But it wasn't just literal interference. Sure, a short, frilly dress makes it hard to crawl through play tunnels. But beyond this concrete type of interference, what may restrict girls even more is their knowledge of how they are *supposed to* move in dresses or skirts. They know that it's not acceptable to kick their legs up high, crawl on the ground, or prop their feet up. There is no real boy equivalent of "sitting like a lady." This same researcher also noted how girls' clothing frequently distracted them from whatever they were doing. Tights had to be pulled up, yanked on, and otherwise adjusted. Teachers interfered with girls' appearance more than they did with boys'. They fixed girls' hair, straightened their outfits, and tightened bows.

Researchers at Kenyon College recently conducted a detailed analysis of dozens of popular Halloween costumes, Valentine's Day cards, and action figures targeted at either girls or boys.[7] They found that 88 percent of the female characters represented in these bits of pop culture wore what the authors called "decorative clothing"—clothes that impede active motion. Only 13 percent of male characters were wearing decorative clothing.

In 2014, Verizon released an online advertisement called "Inspire Her Mind." The video shows a collection of interactions illuminating subtle ways in which parents might discourage young girls from getting or staying interested in science and engineering. One of the snippets in the ad addressed the way girls (but not boys) are discouraged from getting dirty when they play. Verizon was onto something important here. When we dress little girls up in those darling outfits, we're more likely to admonish them not to get their dress dirty when they're playing. But maybe that girl in a pretty dress wanted to examine plant life. Maybe she was learning about insects. Maybe she could have strengthened key motor skills by crawling over or under some obstacles. Instead, she is shackled by her frilly dress and forced to focus on staying pretty in the eyes of others. The way we dress little girls can be the start of a lifelong tendency toward body surveillance. At a very young age, girls learn that their appearance requires monitoring and that how their body looks in clothes is more important than how it moves in clothes.

Does this pattern change when girls grow into women? Most grown women don't find themselves crawling through tunnels. And women scientists aren't told by their colleagues, "Don't get your lab coat dirty." Nonetheless, there are parallels in adulthood. Deborah Rhode, author of *The Beauty Bias*, writes of seeing prominent women professionals made late to meetings because they had to wait in cab lines instead of walking. These women might not be wearing frilly dresses that mustn't get dirty, but the high heels that are such an essential part of a professional woman's wardrobe often make walking to meetings unfeasible.

Moreover, even if the high-heeled professional woman does make it to the meeting on time, she still has to get across the lobby and down the hallway wearing shoes that make walking painful. She will walk more slowly and with less stability than her male counterparts. And now she's in that conference room. And her feet hurt. Every few minutes, a pinching pain in her toes draws her attention away from the meeting for a brief moment. She shifts positions. She briefly disengages from the meeting to fix her skirt and rearrange her legs. This isn't preschool anymore. There's no teacher to monitor how she looks and adjust her dress for her. But that's okay. She's on top of it; she'll do it herself. She's had years of practice. But at what cost to her attention and thinking? At what cost to her psychological health? These are questions we need to be asking.

My students often tease me because I wear cargo pants so often. I can't help it. I just love cargo pants. They have so many useful pockets and they're so easy to move around in. I've been accused of dressing like Kim Possible, which I have decided to take as a compliment, since she spends her time going on interesting adventures. A few years ago, one of my colleagues overheard me in the hallway, extolling the virtues of cargo pants as I opened a package containing a pair I'd just ordered from eBay.

"I don't think women should wear cargo pants," he said.

"What? Why not?" I laughed.

"I just don't," he responded. "They don't look good on women."

"They feel good!" I argued.

"Meh," he replied.

I'm lucky. I work in a job where I can wear pretty much whatever I want. I never have to don the adult woman's equivalent of a frilly dress. But as that little conversation reveals, even in my ivory tower, I can't escape the expectation that my wardrobe should be dictated, at least to some extent, by how much men appreciate it. I still wear cargo pants, though, with a big smile if I see that colleague.

The attention we give our own bodies is directly linked with women's fashions, as is our sense that our bodies must be bullied into an acceptable shape. As Joan Jacobs Brumberg explains in her brilliant historical account, *The Body Project*, as fashions exposed more of the body, they required more body monitoring. Once the legs were exposed, they needed to be smooth and hair-free. As clothing became tighter through the torso (not just the waist), the whole torso needed to become slim. What seemed like greater freedom in terms of clothing actually in some ways translated into more restrictions. Sure, you can bare your arms and legs. You can wear jeans. But now you have to worry about how your arms and legs look. And now those jeans you love reveal the exact shape of your butt. Your freedom to wear a bikini means you have to worry about the size of your thighs.

Chronic body monitoring is a ridiculous price to pay for fashion, but as women, we pay it all the time in dozens of different ways. I don't want young women to feel shame about their bodies. I don't want them to be called sluts when they wear what fashion moguls have decided to be the in style of the season. They should be able to wear whatever they are comfortable wearing. But how comfortable are they? We should have the freedom to dress how we see fit, but we should also have the freedom to be present in the moment. If we are to monitor ourselves, I want us to be able to monitor our thoughts and feelings, our desires and goals, not our appearance.

As I mentioned above, I'm lucky in this domain. I get to spend most of my life wearing exactly what I feel like wearing. Many women don't have that privilege. Yet I still sometimes find myself in shoes that pinch but are *so cute*. Or in a dress that hangs right only if I sit up perfectly straight and keep my shoulders back at all times. Even when we feel we're making the conscious choice to wear clothing that requires body monitoring, I wonder how free that choice is in a culture that practically demands women be sexy in order to be valued. In a world that focused more on what women do and less on how they look, we might find ourselves making different fashion choices.

Worrying About How Your Body Looks Limits How You Can Use Your Body

In a viral video advertisement by the feminine hygiene brand *Always,* a range of girls and women were quizzed about what it means to "run like a girl." I found their reactions heartbreaking. Running like a girl meant showing a lack of physical competence. Running like a girl didn't mean running fast or running hard. It was about how you *looked* when you ran, one more example of girls being taught to be ornaments rather than instruments. I was grateful for this ad, because it fostered important conversations about the disconnect so many girls feel between femininity and bodily competence. Self-objectification is what drives that disconnect. Just as the Stroop task is difficult when you're busy thinking about how you look, when you're focused on the appearance of your body, it's hard to use it as effectively.

The *Journal of Sport & Social Issues* published a unique test of the link between worrying about how your body looks and using your body to do something. Researchers videotaped ten-to seventeen-year-old girls throwing a softball against a wall as hard as they could.[8] Beyond the effects of age and practice that one would expect to see, the girls who reported the highest levels of self-objectification threw the worst. It's hard to act while you're thinking about how you appear.

When you self-objectify, you take an outsider's perspective on your body. Part of the penalty for that outsider's perspective is the loss of what's called *interoceptive awareness*. Interoceptive awareness is a natural sensitivity to stimuli from within your body that send messages about things like when you're hungry, what your heart rate is, or when you need to rest. A recent study out of Kent State University[9] found that the more women self-objectify, the more difficulty they have accurately identifying internal states such as emotions, hunger, and satiety.

Rebecca, our competitive swimmer, learned this difficult lesson firsthand. Working as a lifeguard the summer after her first year in high school, Rebecca found herself with lots of time for reading *Self* magazine when she was sitting by the pool. She recognizes now that "nothing in *Self* magazine was made for people swimming three hours a day," but at the time, she took all the dieting, exercise, and beauty advice she read at face value. Looking back, Rebecca says that magazine was the beginning of her "destruction." Her mental spreadsheet became even more strict, and she added running on top of the training she was already doing as a swimmer. She was on the path to anorexia nervosa, describing her appearance as "bobblehead-esque." By the next school year, she was still on the swim team, but things weren't going so well.

Rebecca explained that her athletic performance was suffering as a result of her obsession with weight. "I was really good my freshman year, really fast. And then I became slower, so I'd work harder, but I'd get even worse. I thought I was doing everything right. I'm working so hard, being healthy. I'm following all the rules. But I had extreme fear of the idea of gaining weight. When I was told, 'You need to gain weight,' I was like, 'No. That's a bad thing. That's a bad thing. Gaining weight is a bad thing.'"

I couldn't believe that Rebecca was left to get diet advice from *Self*. "Did your coach or a health teacher or anyone ever talk to you about what you needed to fuel your body when you're doing a sport?"

Rebecca was clear. *"No. Never."*

I wondered what it felt like for Rebecca when she was training so hard without the fuel she needed for her body. Rebecca told me she had trouble parsing her feelings at the time. "It was hard to tell, to tease apart how you feel when you're training so intensely all the time," she said. "You're always sore and tired. I remember struggling with being cold, like a lot, being wet and cold all the time."

Erin (the art student you met in chapter 3) also reported a direct link between self-objectification and limitations in what she could *do* with her body. After puberty, she became hyperaware of how her body looked when she moved. She knew her breasts jiggled and that embarrassed her. When she and her twin started dressing like boys in order to hide their developing bodies, she remembers that this decision bought them a slightly extended period of childhood. I asked Erin what she got in exchange for deciding to present herself to the world as a boy.

Erin knows exactly what she received in that exchange. She got to think more about what her body could do. She explained, "I felt more comfortable in that way, just using my body. I was a very sprawled out, limby person and very active with my body when I was younger, in that way that only kids are, where you just, like, freely use your body. And I don't think that I stopped that until I was, you know, in junior high and extremely conscious of not being pretty and feeling like I just didn't want to be visible at all."

"Do you think boys get to keep that body freedom?" I asked.

Erin shakes her head in anger. "That's something I talk about with my male friends a lot. I mean, all I have to do is look at a guy sitting on a train. They don't give a fuck about how much room their body takes up or if they're in your way or how they're interacting with you or if they're touching you. They don't care. And, I'm definitely jealous of that."

It's not that Erin wanted to be rude to others, refusing to give way on a crowded sidewalk or insisting on taking up two seats on a packed train. These behaviors just seemed like one more piece of evidence that her body was less free than it should be, and that the limitations she felt had something to do with the fact that she is a woman. In her twenties, Erin started doing Krav Maga, a highly physical type of self-defense training. In some ways, Krav Maga gave her back some of the bodily freedom and sense of safety she lost after puberty. It made her feel like a badass. Maybe she won that freedom back, but it was freedom with an edge of anger and resentment.

Reclaiming Mental Space

Rebecca doesn't struggle to feel bodily freedom in the same way that Erin does. Today she still swims for pleasure and exercise, and feels a freedom from that, from moving her body through the water. But she is still struggling to reclaim brain space. Rebecca spent years during which that calorie counter in her head was taking up a significant amount of mental energy. She estimated it peaked at 80 percent. Today? "Thirty or forty percent," she says. Then she frowns, takes a drink of her tea. "No. Thirty," she decides, emphatic.

Rebecca explains that she still can't quite understand what it would be like to have that mental space back. "I still struggle a lot with comparison and feeling, like, just fascinated by people who don't seem to worry that much about exercise and eating."

"What do you think those women's lives are like?" I asked.

Rebecca seemed sad as she replied. "I mean, a lot more free, you know? A lot less scheduled. I'm in a better place, but it's still a preoccupation. It's still a topic that takes up a lot of space in my mind."

Things are better for Rebecca now. She's happily married to a supportive, accepting man. She enjoys her work and finds it meaningful. But the battle doesn't feel quite over for her, even if she has gained back some of that mental space. To an extent, her beauty sickness has just changed its shape. She finds it popping up in new ways. In particular, though she's not yet thirty, Rebecca has been thinking a lot about aging and how that will change her appearance. She greets these thoughts with a combination of panic and disgust. You can see it on her face—she doesn't want to be thinking about this, but she is. It almost seems like a joke the universe is playing on her. After all those years of frantic calorie counting, now her mind has found something else to worry about. Now there are wrinkles to fear, or graying hair. If Rebecca's friends all have plastic surgery to slow the signs of aging, she wonders if she would be strong enough to opt out. She tells me, "I feel like I'm starting to live with this fear. At times I feel discontent, and then I realize that this could be the best it ever gets, the best I'll ever look is right now. And then I'm, like, 'If you feel discontent with yourself right now, what does that mean for you in the rest of your life?' That's not something I want to deal with."

I can't count how many young women have said something to me along the lines of, "What if this is the best it ever gets?" I'm struck by anger every time I hear it. Beauty sickness has the awful power to double down, taking young women who are already worried about how they look, then telling them it's only going to get worse as they age.

I tell Rebecca, "It's interesting what you said, 'What if this is as good as it gets?' I think about that a lot, about this world where women's power peaks so young, right? You're not even thirty and you already feel like you're on the downhill slope. So what will happen? What will happen if this *is* the best you're ever gonna look?"

Rebecca says she wrestles with the answer to that question. "I don't want to be fixated on my external appearance," she says. "Like, it's a thing, but values wise, that's not what I think is important, so I'm annoyed that it's been creeping up in my mind."

"What do you want to be important as you age?" I ask.

Rebecca sits up straight. "My character and my relationships. What I stand for, and not what I look like."

Ana, our pink-haired high school student, has beliefs that line up well with Rebecca's when it comes to the focus on women's appearance. Ana is angry that our culture sends girls the message that "the only thing you're worth is how sexy your body is or how nice you look." She resents the forces that encourage young women to pick apart the appearance of other women. Ana explains, "There is a cycle of, like, the more you judge other people, the more you feel judged and the more aware you become of yourself."

I asked Ana, "Do you think we'll ever get to a point where it doesn't matter so much how women look?"

Ana is realistic. "I hope, but it takes a lot to change what a society deems as, like, important and valuable." She doesn't actually sound that hopeful.

Ana has three years of high school ahead of her. I ask if she thinks she'll be able to stay true to her values throughout high school, if she'll be able to "stay herself."

Her response is thoughtful and measured. "I hope that I don't change in the sense that I become more enamored with, like, looking a certain way for other people. I hope to stay being confident in myself for me and not for other people."

It might be impossible to quantify the amount of mental resources women lose to body monitoring, but we don't need to see an exact number to know that it's too high. We cannot truly stand up for what we care about if our thoughts are trapped in the mirror. We need to take back some of that mental space for things we care about more than beauty.

5

It's a Shame

A FEW YEARS ago, I gave a presentation on self-objectification to a room full of psychologists. I was discussing how self-objectification leads to high levels of body shame in women and what we might do to reduce it. A colleague interrupted me with a question. "Wait," he said, "isn't it maybe good that women feel shame about their bodies? Maybe you *should* feel bad if you gain weight. It could keep you from gaining weight."

I've heard some variant of that question many times since that day, though rarely so politely. Recently I received a slew of angry emails in response to a *New York Times* op-ed I wrote about Facebook's decision to remove their "feeling fat" emoticon. My basic arguments were that fat talk and body shaming aren't good for women, and that we're wrong when we assume that body shaming motivates healthy behaviors. The authors of these rant-filled emails (all men) took me to task for suggesting that body shame was a bad idea. They proposed that shame is a necessary antidote to the obesity epidemic. One went so far as to tell me that Frenchwomen are all slim (not true, by the way) because French culture so effectively shames fat women. Another suggested that the very future of our country depended on women continuing to feel shame about their bodies. By suggesting that women shouldn't have to feel bad about themselves all the time, I was literally putting our country at risk! I deleted those emails without responding, but if I were more patient and motivated to engage body-shaming trolls, this chapter would be my response to each of them.

We've already established that having your body's appearance on your mind can interfere with cognitive tasks or how you use your body. But self-objectification also packs a hefty punch when it comes to girls' and women's emotional well-being and mental health, primarily because of its power to trigger shame. Here's how the cycle works: You think about how your body looks, which in turn typically makes you think about the body ideal for women. How could it not, when most of us see hundreds of images of this ideal every day? Once that ideal is in your mind's eye, it's hard to avoid comparing your own body to it. And because the beauty ideal is out of reach for almost all women, you're probably going to end up on the losing end of that comparison. That loss, that sense of your appearance not being where it *should be*, is what creates body shame. Beauty sickness is fueled by body shame. The more emphasis you put on your physical appearance, the more shame you tend to feel. This can create a nasty psychological feedback loop, where shame turns your thoughts back to your appearance, which then results in more body shame.

Shame is a complicated emotion. It's a feeling of falling short, of having exposed your flawed self to the judgment-filled eyes of others. It's caught up with cultural norms and social expectations. It leaves you feeling self-conscious and overly tuned in to how others are viewing you. Shame is what Gabrielle feels when she leaves the house without wearing makeup. It's what Artemis feels when she looks in the full-length mirror in her bedroom. It's what Sofia feels when men on the street comment on the size of her ass. If your primary job in the world is to please others with your appearance and you fail to do so, this is a serious failure indeed. The shame that comes along with this failure can devastate even the strongest women.

Across their life-span, women report feeling more frequent and more acute shame than men. When it comes to body shame in particular, the differences are even starker. Research out of the University of Wisconsin found that by age thirteen, body shame is common in girls, and girls already report significantly more body shame than boys.[1] One study published in the *International Journal of Eating Disorders* demonstrated that when college women spent just a few minutes viewing magazine advertisements that featured idealized images of women, their body shame increased.[2] The advertisements increased women's body shame even when the content of the ads was not directly about beauty. Simply seeing an image of the female beauty ideal is enough to remind a woman that she is falling short. A brief glance can be all it takes.

Shame is not the same as guilt. Guilt is generally linked to a specific behavior. It makes you want to apologize, not disappear. Often you can assuage guilt by making amends. Shame is more global; it's deeper and it hurts more. It's not about something you've done, but about who you are at your core. Shame is hard to shake, and it can open the door to anxiety, depression, and eating disorders.

Mary Katherine (who goes by M.K.) is a white forty-six-year-old stay-at-home mom. She knows the feeling of body shame like the back of her hand. Her years of struggle with anxiety, depression, and bulimia are a testament to the power of shame to chisel away at women's well-being. I interviewed M.K. in her posh suburban home, surrounded by framed photos of her four sons (two of whom are older and were away at school). To give us privacy, M.K.'s husband ushered their two young sons into another room when I arrived. The boys' happy screeches as they played were the background soundtrack for our conversation.

Just like Gabrielle, M.K. brought some visual aids for a beauty sickness show-and-tell. Soon after I settled onto one of her comfortable couches, she handed me a memory book from her childhood. This wasn't a regular photo album. Instead, it had a specific page for every year of school. On each of those pages, there was room for a picture and blanks for key bits of information to be filled in. M.K. explained that it's easy to tell by looking at the book when she first started to "have issues." One set of blanks on each page of that memory book asks for height and weight.

M.K.'s kindergarten page was cute but unremarkable. Things got interesting starting with the page for first grade. Below a picture of little M.K. in a puff-sleeved red and white dotted dress, I saw that "50 pounds" had been written in the weight space in pencil. But at some point, young M.K. had messily attempted to erase the 50 and wrote 45 over it. A few grades after that, M.K. had written "80 pounds," but later crossed out the 8 and replaced it with a 7. By the time she was fourteen, M.K. described herself as 5 feet 6 inches and weighing 115 pounds. In parentheses, next to the 115, she wrote in loopy letters, "too much."

"Why do you think you went back and changed that 50 to a 45?" I asked M.K.

"I don't know," M.K. responded. "I mean, I was seven! I must have been getting negative feedback. I was so skinny, but something made me want to be smaller."

"Do you remember going back and changing the numbers?"

M.K. shakes her head and shrugs. "Nuh-uh. No. But I did it. I definitely did it. It was me." She points to her childhood handwriting as evidence.

Beauty sickness is a complicated business, with all sorts of contributing factors. But many of the women I talked to have sharp memories of a specific incident that first solidified their attention to their own appearance. For Rebecca, it was getting her body fat measured in health class. For Erin, it was being told she needed to wear a bra at eight years old. For Artemis, it was a dress that didn't fit. For M.K., it happened when she was in eighth grade.

"I remember exactly what I was wearing. We were on our way to church, my dad was sitting on the couch, and he told me to turn around. And he said he didn't like the way my legs were looking, that my calves looked thick. And that was kind of it, you know what I mean? There was something wrong with me and it was wrong forever after that." Looking back, M.K. describes her legs at the time as "little stick legs." But she didn't wear shorts for that entire summer because she thought her legs were too fat. The obsession that started with her legs grew to encompass her entire body. It lasted for almost two decades.

No matter what M.K. did, the body commentary from her parents continued. "My mom told me that I had the body of a fifty-year-old woman," M.K. recalled, with disgust visible on her face.

You might think I'd stop being surprised by stories like these at some point, but they never seem to lose their ability to leave me openmouthed in shock.

"Your mom told you that?" I asked.

M.K. has thought about the comment so many times, it seems to have lost its psychological impact on her. "Yes," she answered. "She said my face was so fat, who would ever be attracted to me? They were always commenting on what I was eating. I would counter with 'People should like me for what I am. People should like me for what's on the inside.' But it was just this constant barrage of shit about my appearance."

I asked M.K. if she thought her parents made comments like that because they were worried about her health or because they were worried about how she looked. M.K. has no doubt that her parents were focused on her looks, not her health. When I asked her why she thought they were so worried about how she looked, I saw the first flash of anger from M.K. "Oh, god," she began, taking a deep, audible breath, "I think that they were miserable people who didn't think much of themselves, so they didn't think much of their children. And the exterior is what is presented to the world, so regardless of what we were like on the inside, they looked like the good parents with the perfect little family if our exterior looked good to the world. My mom made me get on the scale in front of her when I was fifteen and I weighed 133 and she weighed 136. She got on the scale; then I got on the scale. She made us do it. And she was, like, 'You're fifteen years old. I am forty years old and I've had five kids. What is wrong with you?'"

Not long after that comment from her mom, M.K. began starving herself and abusing laxatives. At one point, she had lost two dozen pounds from her previously healthy frame and looked, as she describes it now, "ridiculously thin." M.K.'s dad, who paid her very little attention in general, finally noticed. He told her she "looked good with all that weight she lost," permanently linking thinness and love in M.K.'s mind.

M.K.'s parents are both still alive today, and M.K. says her "Catholic guilt" keeps her interacting with them despite the psychological wounds they've left her with. Her dad is sick now, suffering from advanced Parkinson's. He was angry at M.K. recently because she had cut a visit with him short. M.K. asked him to forgive her, but he refused. M.K. lost her temper.

"I forgive you for things!" she exclaimed.

"What?" her dad asked, having no idea what she would need to forgive him for.

M.K. told me everything came spilling out after that. "You said I was fat when I was a teenager!" she cried.

"You were," her father responded, deadpan.

That comment sent M.K. into a tailspin. She came back home, binged for two weeks, and gained ten pounds. "Two words from my father throws me right back into that," she explained.

The Unique Cultural Shame of Fatness

For women, the cultural beauty ideal has varied a bit in recent years in terms of features like breast size, height, or hair color. But the one constant is thinness. Even women who are held up as examples of a "curvier" body ideal typically still have flat stomachs and no visible cellulite. Deviations from the thin ideal are often met with hostility and ridicule.

Consider the controversy that erupted over the 2014 Scooby-Doo movie, *Frankencreepy*. In the film, the typically svelte Daphne is struck by a terrible curse—she's magically turned from a size 2 to a size 8. Let's set aside the fact that a size 8 isn't fat, and that the drawing of "fat Daphne" looked nothing like a size 8 in the first place. There's something even more pernicious at work here. Daphne's curse was punishment for admiring the shape of her size 2 body. It was her punishment for vanity. Lessons for young women in this film?

1. You're pretty only if you're a size 2.
2. Don't admire that size 2 body or feel good about it, or you'll be punished for your vanity. Other people are the only ones who can approve of your body. Never you.
3. If you succumb to vanity, you'll be punished in the worst way imaginable for a woman. You will gain weight.

The weight you gain will make you look like (as writer Tom Burns of *The Good Men Project* put it) "Violet Beauregarde from *Willy Wonka*," even though being a size 8 would actually leave you thinner than the average American woman.

Children in our culture learn early to associate body fat with a range of negative characteristics. In a study out of Williams College, researchers told stories to children aged three to five, in which one child was mean to another child.[3] Afterward, the researchers showed the kids pictures of other children who ranged from thin to chubby and asked which child was the mean one. The children in the study assumed the chubby child was the mean child. They were also less likely to say they'd want to play with a chubby child. Overweight children in the study were *even more* likely to link chubby and mean, a worrisome finding suggesting that at such a young age, these kids had already internalized a terrible lesson about their own self-worth.

Parents often unwittingly reinforce the notion that body fat is something to be ashamed of when they criticize their own bodies or others' bodies in front of their children. In a study of nine-year-old girls published by the journal *Obesity Research*, scientists found that children whose parents focused on body shape and weight loss were more likely to stereotype fat people.[4] These stereotypes lay the foundation for bias and discrimination in all kinds of real-world settings.

Weight-related bias is widespread, and research consistently finds that women are the targets of this type of bias much more frequently than men. A study out of Michigan State University's School of Human Resources and Labor Relations found that women are sixteen times more likely than men to report weight-based discrimination in the workplace.[5] Although men and women both pay a price in the workplace if they are perceived as fat, the price women pay is much steeper.

Researchers often examine weight-related prejudice by constructing hiring scenarios and providing information about hypothetical applicants' weight or body size. Scientists at the University of Wisconsin and University of Northern Iowa found that both men and women indicate less desire to work with a fat person (compared to a thin person), but particularly if that fat person is a woman.[6] Compared to fat male applicants, fat female applicants are also less likely to be recommended as new hires.[7] A study out of the University of Missouri using similar methodology found that overweight political candidates are penalized for their weight *only* if they are women.[8] The same pattern is also visible in studies of romantic relationships. In a study of college students at Indiana State University, heavier women were significantly less likely to date compared to their average-weight peers, but men's weight had no impact on whether they dated.[9]

Particularly for women, a failure to be thin is viewed as a deep character flaw, indicating laziness, gluttony, or lack of discipline. When M.K. eventually built up the courage to tell her parents she had an eating disorder, her dad's first response was "Why can't you control what's in your refrigerator?"—as though she just wasn't *trying* hard enough.

One University of New Mexico professor captured these types of cruel beliefs perfectly when he tweeted, "Dear obese PhD applicants: if you didn't have the willpower to stop eating carbs, you won't have the willpower to do a dissertation #truth." In addition to being asinine, the tweet is factually incorrect, as beautifully demonstrated by the woman who created a Tumblr called "Fuck yeah! Fat PhDs!" where dozens of women shy of the thin body ideal proudly noted their graduate degrees. Of course, given our culture's relentless focus on women's bodies, even being thin isn't a guarantee that you won't be body shamed. Plenty of people are happy to shame women for being too thin as well, so narrow is the range in which women's bodies are acceptable. Nonetheless, the women who get hit the hardest with this type of shaming are those who are heavier than our culture's rigid body ideal for women.

There Is No Good Reason to Body-Shame Women

Those pro-body-shaming emails I mentioned at the beginning of this chapter might make more sense if there was any evidence to support the claim that body shame alters your behavior in beneficial ways. If you're not being the person you want to be, shame could, *in theory*, alert you to that fact and help you reconsider your behaviors. In a situation like that, shame would be adaptive. But the process is different for women's body shame in three key ways.

First, the beauty ideal is unattainable, so it's neither fair nor sensible to expect someone to feel shame for not attaining it. You're not a failure; the system is rigged. Second, body shame generally does not push women any closer to the beauty ideal. In fact, it often pushes them further from it. Third, getting closer to the beauty ideal does not guarantee health, and in some cases may in fact make you *less* healthy.

I'll explore the first point, about the unattainability of the female beauty ideal, in more detail in chapter 7. So for now let's focus in on the other two points. Because even if you think the beauty ideal *is* achievable, shame is a disastrous way to go about trying to get there. That shame you feel when you fall short of this particular ideal isn't helpful. It only hurts.

Shame Is Not a Diet Plan

Though M.K. says she's "come a long way" in terms of body image, she also told me that there has never been a day when she felt truly comfortable in her body. Some days are worse than others. On the bad days, she catches a glimpse of her naked body in the mirror as she exits the shower and is greeted with a wave of body shame. Her go-to response to body shame is bingeing. As a result of cycles of bingeing and purging alternating with extreme diets, M.K.'s weight has fluctuated widely, within a range of nearly one hundred pounds. For years she would try to "erase" the shame of bingeing by abusing laxatives. But M.K. stopped taking laxatives when she was twenty-five, afraid that the damage they were doing to her body might prevent her from being able to have children. There was never a time when M.K.'s body shame motivated her to make healthy choices for her body. Shame is not a diet plan.

Let's start by taking apart the idea that body shame provides a pathway to the ultrathin body ideal for women. Of all the research conducted on obesity, there is not one drop of evidence that fat shaming helps to move people toward thinness. In fact, the opposite is true.

A research study in the *Journal of Health Psychology* showed that the more young women receive negative comments about their weight, the *less* likely they are to exercise.[10] In a different study of over five thousand adults across the United States, results demonstrated that experiencing weight stigmatization and discrimination is associated with an *increased* likelihood of overeating and more frequent consumption of convenience foods.[11] These same patterns show up when you study children. Several independent labs have confirmed that among children, being teased about weight predicts increased binge eating.[12] Just like adults who experience weight stigmatization, kids who are teased about their weight show less interest in sports and less physical activity overall. Similarly, a 2016 study of over 2,000 adolescents in Minnesota showed that teens who felt worse about their bodies exercised less and were less likely to eat fruits and vegetables.[13]

It's not uncommon for women to make grand plans when they're in the midst of a crisis of body shame. Perhaps they were just teased for their weight or rejected by a potential romantic partner. Or maybe those jeans that used to fit can't be buttoned anymore. The first response might be "I'm going on a diet. I'm cutting out carbs. I'll go to the gym every day. No more dessert." There are two problems with reactions like that. First, such reactions tend not to lead to any long-term healthy changes in behavior. Second, they may lead to behaviors that start women down the dangerous road to disordered eating, because when you feel yourself falling short of the body standard, too often the reaction is to do absolutely anything you can to move yourself closer to that skinny ideal.

Eating disorders are serious business. Anorexia has the highest mortality rate of any disorder listed in the *Diagnostic and Statistical Manual of Mental Disorders* (DSM), the psychiatric guidebook.[14] The short- and long-term health consequences of anorexia and bulimia are numerous (dental erosion, esophageal or stomach rupture, electrolyte disturbances, arrhythmias, and loss of bone density, just to name a handful). Bulimic behaviors are shockingly common among young women. A 2009 University of Minnesota study found that almost half of college women reported engaging in bingeing or purging at least once per week.[15] I've never made it through a school year without being approached by multiple students who have concerns about friends engaging in these types of behaviors. Occasional bingeing and purging and on and off starvation diets have almost become the norm, encouraged via a swath of advertising telling us simultaneously that we should indulge our desires *and* shrink our bodies.

When we're in emotional distress, we will usually take action to try to make ourselves feel better, even if that means trading a short-term mood boost for less appealing long-term consequences. Feeling better may come in the form of a pint of ice cream or a bag of chips that relieves your emotional distress temporarily, but triggers a spiral of shame. You felt bad about your body, so you ate something to feel better, and now you feel even worse about your body. This is one of the routes through which experiencing weight stigmatization can lead directly to binge eating.

Psychologists have long known that shame can trigger bulimic behavior. In one study published in the *International Journal of Eating Disorders*, researchers showed a group of women struggling with bulimia slides of their favorite foods.[16] Next, the women were asked to spend a few minutes remembering a time they felt sad, while experimenters played somber music in the background. This is called a mood-induction paradigm—the goal was to put these women in a negative mood. Once the women felt sad, researchers showed slides of their favorite foods again. When they were sad, the women paid even more attention to the images of food and reported greater craving. In other words, emotional distress can make it harder to maintain healthy eating habits.

I asked M.K. what she thought about this "body shame equals diet plan" idea I so often hear. Consistent with the research just reviewed, she links body shame directly to bingeing, describing body shame as the "worst way to try to lose weight." M.K. finds that her first instinct when she feels shamed is to eat—not to go for a walk or cook a healthy meal. When she is ashamed, it feels like there's no reason to take care of her body. "When you feel ashamed and depressed," M.K. explained, "you're not bopping out of bed in the morning and taking a walk around the block and eating oatmeal with blueberries on the side. You're going to Dunkin' Donuts, because you feel like there's no hope. Shame has no tie to hope at all."

To be fair, not everyone binges when they're upset about their body shape. Some take another route. They decide to just stop eating for a while, to skip meals, or to cut down their caloric intake drastically. Paradoxically, this often makes dieters *more* obsessed with food. When you're hungry, it's hard to focus on anything else. Think of the double whammy this is. Your attentional resources are already limited because of body monitoring, and then hunger gets added to the mix, providing even more distraction.

One thirty-six-year-old woman I spoke to estimated that she had been on a diet for somewhere between a third and a half of her life. She explained that one of the worst parts of dieting was how it made her spend more time thinking about food. Not someone who enjoys cooking, she told me, "I feel like I spend almost all of my time planning and supporting food prep. I sometimes think about what I'm going to eat when I should be working. If I have an event, I worry about what I will eat there."

M.K.'s weight-loss distractions were even more extreme. When she was in high school, her parents had no idea she had an eating disorder. M.K. would time her laxative abuse so that her multiple painful trips to the bathroom would be in the middle of the night. But she explained to me that the timing was never an exact science. The next day in school, she'd be unable to concentrate, living in fear of having an "accident," monitoring her body for telltale cramping or gurgling. Most people experience cravings for foods, but one of the factors that distinguishes those who crave and eat a reasonable amount from those who crave and binge is that those who binge are more likely to first engage in dietary restraint. In other words, drastic cutbacks in calories increase the likelihood of later bingeing. This can be due to the frustration associated with restraint, the way restraint makes you hypersensitive to food, or even metabolic changes associated with dieting. On top of the distraction of hunger and cravings, these types of drastic reductions in calories are associated with mood swings. And if mood swings weren't enough to contend with, research from the British Institute of Food Research found that those on very low calorie diets also show impaired cognitive performance, scoring lower on memory, attention, and reaction time tasks compared to women who aren't dieting.[17]

What dieting *doesn't* seem to do is lead to permanent long-term changes in body weight.[18] Authors of a UCLA-led review of long-term outcomes of calorie-restrictive diets concluded that diets tend to be unsustainable and do not consistently lead to improved health outcomes.[19] Across dozens of published studies, the authors found that very few people maintain weight loss. Most quickly regain lost weight, and many gain back more weight than they lost through dieting.

Responding to a kerfuffle over having her body airbrushed thinner on the cover of *Flare*, actress Jennifer Lawrence recently called out those who fat shame. In an interview with Yahoo! CEO Marissa Mayer, Lawrence asked, "What are you going to do? Be hungry every single day to make other people happy?" Careful reviews of weight-loss research suggest that for most people, the only way to permanently diet your way to a significantly lower weight is to be willing to remain hungry every single day.

We Too Easily Confuse Thinness with Health

Even if the female body ideal were attainable and even if shaming women helped them attain that ideal, doing so still wouldn't actually be a good way to make women healthier. You absolutely do not need to have visible ribs, a thigh gap, or a perfectly flat stomach in order to be healthy. You're kidding yourself if you think you can always tell by looking whether someone is healthy. In a study of over 5,000 adults led by the Albert Einstein College of Medicine, researchers came to a stark conclusion: in the United States there are many "normal-weight" people (around 24 percent) who show poor cardiovascular and metabolic health, and there are many obese individuals (around 32 percent) who are metabolically healthy.[20] Medical ethicists have made strong arguments in recent years that physicians should *not* focus on weight loss when treating obese patients. That focus too often exacerbates the shame patients already feel and keeps people from seeking needed medical care. Instead, doctors should focus on increasing direct indicators of physical and psychological well-being in their patients. This recommendation to move the focus from an appearance-related variable such as weight to more pure indicators of health is consistent with everything we know about the dangers of self-objectification.

Thanks in part to the pervasive nature of our beauty-sick culture, much of the purported emphasis on "health" is often a thinly veiled concern about aesthetics. In 2015, researchers from Appalachian State University and Kent State University published an analysis of cover captions from *Women's Health* and *Men's Health* between 2006 and 2011 (108 covers in all).[21] Among the most prominent captions on the covers of *Women's Health* (i.e., those with the biggest font), 83 percent were framed in terms of appearance and/or weight loss. *Women's Health* captions were also more likely than *Men's Health* captions to emphasize appearance. None of the prominent captions for *either* magazine focused directly on health, despite the fact that the word *health* is in the titles of both.

The research above makes it clear that you can't tell whether someone is healthy just based on their body weight, but even if you could, quite frankly, that's no excuse to treat anyone badly. "Health" should never be a prerequisite for being loved or being treated with dignity and respect. Women who don't meet our culture's rigid beauty ideals don't owe the world some demonstration of their metabolic rate or cardiovascular fitness in order to be treated well or to prove that they're "just as good as thin women."

You can see why I felt so incredulous when I received emails from strangers arguing that we need more fat shaming in this culture. Both weight-based discrimination and obesity have continued to increase over time—there's no sign that one stops the other. I'm deeply skeptical of those who claim they're trying to "help" women by shaming them for their body size or shape, or those who say they fat-shame because they are worried about women's health. Don't imagine for a moment that any woman in this culture who struggles with weight is under any illusions about what her body looks like compared to the ideal. There is never a need to point out this gap. You are not doing her a favor. She already knows, trust me. Given the rampant fat shaming in this country, how could anyone imagine that obesity is a result of the fact that we simply don't make people who are fat feel bad enough about themselves? Please.

For those who claim to engage in fat shaming because they are concerned about women's health, I humbly suggest doing something that actually helps women's health instead. Volunteer at a local clinic. Donate to Doctors Without Borders. Support research and legislation that protects women's health. Cruelty is not a health intervention. It's nothing more than a misguided, self-righteous attempt to boost one's own self-esteem.

Here's what I have to say to everyone who seems to believe that we should encourage women to feel body shame in order to promote weight loss. Even if you're not convinced by all the empirical data reviewed above, why would you ever want to employ a health intervention focused not on caring for one's body and treating it well, but rather based on loathing your body? Why would you want women to hate such an intimate and important part of themselves? What we need instead is to feel so at home and comfortable in our bodies that taking care of them feels natural and automatic. You don't take care of things you hate.

The Depressing Reality of Body Shame

Body shaming doesn't move women closer to an unrealistic body ideal, and even if it did, that body ideal should not be conflated with health. But there's an additional reason to be concerned about the pain of body shame. Shame is highly associated both with everyday depressed moods and with more serious clinical depression. We can't ignore this important outcome of body monitoring.

Scientists have been studying depression long enough to know that we can't pinpoint one single cause. Hormones matter, genetics matter, life events matter, and your temperament and ways of thinking matter. But the shift in depression that occurs in adolescence seems to have something to do with how (and how much) teenage girls think about their bodies. A ten-year study of over 1,000 children found that girls and boys begin junior high with similar rates of depression, but by the time they are fifteen, more than twice as many girls than boys are depressed.[22] Importantly, a University of Wisconsin study of almost 300 girls showed that the gap between girls and boys in self-objectification shows up *before* the gap in depression.[23] That's a good piece of evidence that the increases in body monitoring typically accompanying puberty should be considered a key risk factor for the development of depression in girls and young women. One study out of Northern Illinois University and the University of North Dakota found that girls with more negative feelings about their body were more likely to think about suicide.[24] Body image was an even stronger predictor of suicidal behavior than other risk factors like feelings of hopelessness or depression.

In college women, self-objectification is associated with all types of negative moods[25] and is directly related to symptoms of depression.[26] A study led by the University of Toronto followed adolescent girls for five years, and found that if girls engaged in less self-objectification and body monitoring, rates of depression went down.[27] In other words, if we can reduce the amount of attention young women are dedicating to the appearance of their bodies, we might decrease rates of depression as well.

This link between depression and monitoring your appearance probably makes sense to most of us. Self-objectification promotes rumination—the tendency to focus on your own psychological distress. Rumination is a repetitive type of thinking, where the same negative thoughts and scenarios play in a loop that you can't seem to interrupt. It's an extremely strong predictor of depression, and women ruminate more than men. But while there are differences between men and women in overall rumination, there is an especially big gender gap when it comes to rumination about appearance and body image. Too many women are experts at this type of rumination.

When I think back to that first psychotherapy client I had, I'm disappointed that I didn't know enough at the time to follow up more meaningfully on her feeling of being looked at. Instead, I focused directly on her symptoms of depression, without realizing that the way she experienced her body might have been feeding that particular monster. I'll never know now, but it's possible that targeting that young woman's body shame might have been a helpful way to fight her depression. For M.K., the body monitoring that started when she was in junior high with an offhand shaming comment about the size of her legs has fanned the flames of eating disorders, depression, and anxiety for years. M.K. used the term *brain space* when talking about the impact beauty sickness has had on her mental resources. I like that term. It perfectly captures the notion of limited cognitive resources and the trade-offs we make when so much of our brainpower is dedicated to thinking about how we look.

M.K. told me about a full-length mirror in her bedroom. She catches herself in that mirror every time she sits to put her shoes on in the morning. Despite the progress she's made in accepting her body, glancing at that mirror often still leads to her cataloguing body areas she'd like to change. When I asked M.K. if she'd ever thought of tearing that mirror down, she told me she feels she's made enough progress to be able to leave it up.

"I don't think I need to," she explained. "It used to be, I'd say, ninety percent of my thoughts were about being fat, and now maybe it's ten percent?" She's not certain of the number, but M.K. knows that she has reclaimed a substantial proportion of that stolen brain space.

"How many years of your life did you spend in that ninety percent zone?" I asked.

M.K. tilts her head up to the ceiling, running through a mental calculation. "Oh, god," she says, shocked at the answer she comes up with. "Um, fifteen?"

I ask M.K. to imagine how her life might have been altered if she hadn't sacrificed 90 percent of her brain space to body monitoring for so many years. "How would your life have been different if you had that brain space back?" I asked. "What would you have done?"

I could see it in M.K.'s body language and hear it in her voice. It was excruciating for her to think of this road not traveled. "I would have had self-esteem, which would have changed everything in my entire life."

She continued, "I would actually have thought something of myself instead of nothing of myself. I think that I would have been very different in relationships with men."

"How do you think you would have been different in your relationships with men?" I asked.

M.K. sounds disgusted. "I used to sleep with people just because I wanted them to like me. I didn't value my body, so I just gave it away. I never thought anyone would like me enough to wait. And that's the saddest thing in the world, thinking no one would ever like me enough to wait to have sex with me, to wait to get to know me. So I just gave it away, you know?"

M.K winced as she finished. "If I thought more of myself, I think, I just can't tell you how much I wish I could be me now and go back. Just to be able to go back."

M.K. can't go back, of course. But it is clear that she has decided to no longer let her life be held hostage by beauty sickness. M.K. explains: "You can do Botox all day long, but you're still going to have wrinkles and you're still gonna get gray hair. You can't stop it. You know, someday I'm going to look in the mirror and have flesh hanging off me, but that's what aging is. I mean, the bottom line is that I just want to be comfortable in my body now."

So M.K. is still fighting, but winning more battles than she loses, making her experiences both a cautionary tale and a tale of hope. Her experience shows us the costs of self-objectification. She teaches us that we must take all of this body monitoring seriously because it has serious consequences. It leads to paralyzing body shame. It makes exercise a burden and eating a battle. It's an important precursor to serious depression in young women and is even associated with suicidal ideation. It makes us beauty sick.

But beauty sickness isn't just a force that attacks women's mental and emotional well-being. It's a literal thief as well, all too often walking away with two of women's most precious resources: time and money.

6

Your Money and
Your Time

MY FRIEND BILL is a lover of statistics. It's not uncommon for him to pop into my office, gleeful about some new finding or technique. One day he came waving a piece of paper, excited about findings he generated using a technique called *dustbowl empiricism*. I'll spare you the statistical details and put it in simple terms. What Bill did was examine around 1,000 statements from personality tests that were completed by over 100,000 participants around the world. He used a program to find the statements that could best predict whether the person taking the survey identified as a man or a woman. The term dustbowl empiricism means that he wasn't interested in *why* any given item correlated with the test taker's gender, just whether and to what extent it did. It's like going hunting with no specific animal in mind. You just see what you run into. What Bill ran into was a finding that elegantly captured how the cost of beauty varies by gender.

Bill jabbed his finger at the piece of paper, directing my attention to one of the items that most accurately identified the gender of the survey taker: *"I have spent more than an hour thinking about what to wear."* Women were substantially more likely to agree with this statement than men. "Look at this," he said. "I thought you'd be interested."

Of all the psychological traits and interests that might distinguish men and women, how much time one spends thinking about what to wear was one of the *most* distinguishing characteristics. To be clear, this finding doesn't mean that *all* women spend tons of time thinking about their wardrobe or that men never worry about what to wear. Of course not. But it is one more powerful piece of evidence illuminating how much women's lives are shaped by beauty concerns.

I sighed when Bill showed me that finding. "That is interesting. And pretty upsetting," I said. Bill agreed. Then he walked down the hall, heading back to his data analysis, and I returned to thinking about the price of beauty sickness. In addition to paying a price in terms of reductions in bodily freedom, cognitive resources, and mental and physical health, women actually, literally, pay for the malaise of beauty sickness with their own money and time. We pay and pay.

JESS*, A WHITE THIRTY-THREE-YEAR-OLD DIRECTOR at a midsize technology company in Minnesota, sneaked away from work on her lunch break to meet me at a nearby diner and discuss the costs of beauty sickness. Our waitress kindly tucked us away at a little table by the window, where we watched a light snow fall as we talked and ate waffles and chicken fingers. Jess has thought a lot about the costs of "keeping up appearances" as a woman. She spent five years in the high-pressure world of a top consulting firm, where looks always seemed to matter more than they should have. Even now Jess finds herself spending more time on her appearance than she would like to. She recently returned to work after having her first child, so every minute feels particularly precious to her these days. She turned her cell phone to show me a picture of a happy little baby.

Jess was slower than most to enter the everyday beauty pageant so many women face. She grew up in a tiny blue-collar town. The nearest mall was several hours away, and Internet shopping had not yet reached its heyday. Fashion-focused outings were few and far between. Jess was more focused on academic achievement anyway, and her hard work earned her a scholarship to a prestigious women's college.

Jess loved the environment of her women's college, and to some extent, it gave her freedom from having to think much about how she looked. The *male gaze* we talked about in chapter 3 was simply absent from many of her everyday experiences. But once in a while, her small-town roots would leave Jess feeling lost when it came to the beauty knowledge so many other women just seemed to automatically possess. That emotion hit her hard when she attended a party hosted by her running coach. There were students there from other schools, men and women, and they all knew it was a big deal to be invited to this coach's home. Jess described how she prepared to attend the party.

"So I'm, like, 'Okay, well, I will wear my nicest khakis and my nicest sweater.' When I got to the party, it was the first time I realized, 'Oh my god, I don't have a dress.' Like, all the other women were in a skirt or a dress. So it very much dawned on me that I didn't know how to do that, how to dress the same way."

Jess can still connect to the feeling of embarrassment that resulted from that realization. "I just . . . I felt like I didn't fit in." She shakes her head, remembering that night. "And I felt kind of ashamed at what I had picked to wear, and stupid for thinking it was a nice outfit, and just very uncomfortable. It made it difficult to enjoy the party because the whole time I was thinking, 'Oh, I don't really fit in with what I'm wearing, I made the wrong decision.' I legitimately thought it was a nice outfit, and I was so off and wrong. I thought, 'What is wrong with me that I don't understand this?'"

When I asked Jess if she remembered what the guys at that party were wearing, she laughs. "Jeans and button-up shirts. You know, the uniform of men since whenever blue jeans and button-up shirts were invented. I'm sure that they probably didn't even think about it. They probably just wore it."

I had a pretty strong emotional reaction to Jess's story. I still regularly find myself in settings where I feel the same way she did that night. I too often feel I missed some sort of orientation, a woman-specific training on what to wear, when, and how. I'm sure lots of other women are in the same boat. But what I most like about Jess's story is how clear she is about the consequences of that feeling of not looking quite right. That feeling stole her ability to enjoy what should have been a really fun event for her. As these types of experiences accumulated, Jess felt she had to spend more time and money getting the "right look."

Jess began interviewing for consulting jobs as she neared her college graduation. Having the right clothes, jewelry, hair, and makeup started to seem even more essential, but she still felt somewhat at a loss. She knew when women "looked nice," but didn't necessarily feel she knew how to imitate that look.

Just as at that party, interviews seemed like another occasion where men just didn't have to make that many choices. They didn't have to be so aware of how they looked or work so hard just to make sure they fit in. Jess catalogued the list of things women have to think about. "How do you do your hair? Do you curl it? Do you leave it straight? Do you pull it back? How much makeup is too much makeup and what is just the right amount? Dangly earrings or pearls? Skirt or pants? Heels or flats?"

Men in Jess's field simply don't have to make this same number of decisions. As Jess explained it, they basically just choose between wearing their suit jacket or not. "There aren't very many ways to mess up," she says, rolling her eyes. "Women don't have that equivalent of a uniform."

I asked Jess if she would like it if women had a workplace "uniform" similar to men's. "I would love it!" she exclaimed. "It would take so much of the burden away."

Once she was hired as a consultant, the expectations for Jess to manage her appearance were ratcheted up even more. Her firm, like many, was very image conscious. Part of the training for this firm was an etiquette class. Although both men and women attended the class, the focus felt tilted toward women. After all, men's suits may be expensive, but they're pretty straightforward. And removing the jacket and/or tie is a fairly basic modification for less formal settings. On top of that, a man doesn't need to buy very many suits or pairs of shoes to have a complete work wardrobe. Things aren't so simple for women.

The etiquette teacher gave guidelines for what the women could wear to work. Jess explained, "You can't just wear clothes. She said you have to wear an outfit."

I was honest with Jess that I had no idea what that meant. What makes something an outfit? "Right," Jess began, easily parroting what she'd been taught. "An outfit for women is always composed of at least three elements. Pants and a shirt is not an outfit. If you add an interesting necklace or a nice-looking belt, that's an outfit."

I glanced down at my jeans and long-sleeved T-shirt. Not an outfit. "Three elements is more expensive than two elements," I pointed out to Jess.

"Three is a lot more expensive!" she confirmed.

I asked Jess if she was currently wearing an outfit and she indicated she was, swinging her legs out from under the table to show me. She was wearing a dark-colored skirt and a white sweater, but it counted as an outfit, she explained, because she paired the skirt with patterned tights.

Jess has to spend a decent amount of time thinking about what's in her wardrobe and how everything might fit together. "Do you enjoy that," I asked. "Or is it a burden?"

Jess answers quickly. "I hated it. I still hate it." She uses Stitch Fix now, because they send her outfits and tell her how to put them together. But she doesn't derive any pleasure from it.

"It's an expensive and complicated standard to meet," I said.

"It is. Yeah," Jess agreed. "Especially when you add in purses and jewelry and nails. You can't just carry some ratty purse."

All this talk made me grateful not to work in a setting that requires so much in terms of beauty time and money. I glanced down at my own not-far-from-ratty purse, which I had carelessly tossed under the table. It was sitting in a small, dirty pool of melted snow.

Beauty Doesn't Come Cheap

Although the wage gap between women and men has narrowed since the 1970s, women working full-time still earn substantially less than men working full-time. Likewise, women are more likely to be found in lower-paying occupations and industries. Women are also less likely than men to receive employer-provided health insurance or retirement savings plans, both of which have a serious impact on financial well-being. Consistent with all of these findings, women are much more likely to live in poverty than men. Economists and policy makers can argue all day about the reasons for these gaps (discrimination, different career interests, child care and family responsibilities), but regardless of the origin of these differences, it's a basic truth that on average, women bring home less money than men.

While many women (and some men) rail against the gendered pay gap, too often, we don't consider the gender gap in time and money spent on beauty. We tend to see concerns about our appearance as separate from career concerns, or imagine that the problems we see in the mirror have no impact on problems we hope to address in the world. But time and money matter. They're essential sources of power and influence and also major sources of freedom. We have no problem talking about some types of costs and how they matter—health care, child care, housing. There's no reason we shouldn't also give careful consideration to the cost of beauty.

Let me be very clear: I'm not arguing that women have any moral obligation to give up all of their beauty spending and practices. In additional to being wildly unrealistic for women in many settings, it's also not something most women would want to do. I'm one of those women. For example, I *like* lipstick. I don't want to stop buying or wearing lipstick, even if I also don't want to be *required* (either officially or implicitly) to wear lipstick. But just like most human behaviors, our beauty choices are on a continuum. Unless we think carefully about the consequences of our individual beauty choices, how will we even know if they're the best choices for us? The risk is in *not* thinking about the cost of beauty, in letting these costs become an integral part of your life without ever truly deciding what you do or do not want to spend time and money on.

A few years ago, one of my dear friends, a language professor, was arranging to go with a group of her students to a local shop that did eyebrow shaping. It was a bonding event for this particular group of young women, and they liked the idea of having some women professors along for the ride. A group of us were eating to-gether at a long table in the dining hall, and my friend invited me to join the eyebrow adventure. I thanked her but declined. "I don't want another 'beauty thing' on my list," I explained. Things got a little bit heated at our table after that. The students really wanted this outing, she responded; it wasn't like they were being coerced into getting their eyebrows done. And it wasn't expensive. Only ten dollars. Why wouldn't I go? I stuck to my guns, but I worry I came off as unintentionally judgmental.

The truth is that I had no problem with the outing and, in fact, was pretty convinced I would have fun if I went. There was no part of me that wanted to shame young women for desiring this type of get-together. I'm a firm believer that we need less policing of other women's behavior, not more. This wasn't about me telling young women they shouldn't get their eyebrows done. It was about me making a conscious, deliberate choice for myself.

At the time, I honestly had no idea what brow shaping even en-tailed. It wasn't that I was worried I wouldn't like it. Quite the oppo-site. I was worried I *would* like it. And if I did like it, if I felt my face looked better after having it done, brow shaping would become one more thing I felt like I needed to do on a regular basis just for basic maintenance of my appearance. I didn't want any more things on that list. I didn't want to feed the part of me that already monitors my appearance in a dozen different ways. That part of me is busy enough trying to get me to make a haircut appointment on a regular basis. It doesn't need to worry about my eyebrows too. Other women might welcome having a new beauty ritual. They might have the time and money to incorporate something new, and might experi-ence the whole event in a positive way. And that is absolutely fine, of course. But I knew it was the wrong choice for me at the time.

Just as I think it's wrong to criticize women for the time and money they do spend on beauty, I think it's wrong to criticize women who opt out of this spending in different ways. These two criticisms are just different sides of the same useless coin. I'll never forget the time I was in the local YMCA locker room and overheard a young woman say to her friend, "Look at those dirty heathens. I can't believe they don't wax." She was reacting to some middle-aged women who had chosen not to wax or laser off most of their pubic hair, and was clearly suggesting that to opt out of that practice was somehow shameful. If we shift our culture, even just a little, so that we all spend less time talking and thinking about women's looks, all women will be freer to make choices about how much time and money they want to invest in beauty.

Jess says she has caught herself falling into the trap of making snap judgments based on what other women in her workplace are wearing. I asked her for an example. "Well," she began, "there's this one woman . . ."

"You have a guilty smile!" I laughed.

"Yeah, it's such a guilty thing," Jess confirmed. "Because it's not nice. This woman, she's constantly put together. But yesterday she wore leggings, boots, and a company T-shirt."

When Jess saw that outfit, she caught herself thinking, "What is she *doing*?" But then Jess stopped, reminding herself that that woman might have had a difficult morning. She might have needed to put on whatever she could just to get out the door that day. "It's totally rude that I judged her for it," Jess said, "but at the same time, I totally noticed." It's a basic truth that how we present ourselves, especially in the workplace, has an impact on how others treat us. What I'm advocating is not a beauty-spending moratorium, just an honest accounting and review. I'm not suggesting that we blind ourselves to the real impact of grooming on our everyday interactions. There can be substantive costs to opting out of certain beauty practices. But once again, this issue isn't black and white. It's worth considering whether we might shift *some* of the time and money we spend on beauty to other endeavors, and what impact this might have on women's lives. We also need an honest discussion about the actual costs of some beauty practices and what trade-offs we make in order to meet these costs.

At Northwestern, I teach a course entitled "The Psychology of Beauty." For one class activity, I give students sheets on which to record how much money they spend on various beauty pursuits. I give them an exemption for clothing, because it's hard to quantify what counts as essential in terms of one's wardrobe and what counts as a beauty-related purchase. I also tell students they don't need to include gym memberships or fitness class fees, since it can be hard to draw the line between working out to change one's appearance and working out to increase health. Finally, I instruct my students to skip hygiene basics—soap and toothpaste don't count, but purchases of tooth-whitening supplies and special beauty soaps do. I encourage my students to think of every other beauty expense they can come up with. Makeup, manicures, haircuts, highlights, styling products, nonmedical dermatology, waxing, razor blades, shaving cream, lotions. All of it. After being given these instructions, my students make their lists. They include all the beauty-related things they spend money on and indicate about how much each of these things costs and how many times per year they have to buy it.

Try it yourself at home. It can be pretty eye-opening. I complete the exercise along with my students every year, and am always unpleasantly surprised at my own total.

My students' estimates of their beauty expenses vary widely. One young woman totaled up close to $5,000 a year. Remember, these are full-time students. They're not out in the world earn-ing full-time salaries. One young man estimated he spent $50 per year. He cuts his own hair and gets his shampoo, razors, and shaving cream from the free samples given out in the student center, he explained.

My best guess is that all of these students were underestimating. It's hard to think of everything you spend money on. A more accurate accounting could be done by keeping a diary of everyday purchases and expenses for a year (or, at the very least, by going through your bathroom cabinets and drawers while you do a one-time accounting). But regardless of how imprecise these estimates may have been, many women were shocked at the amounts they came up with, and at least a little irritated by how much lower the men's totals were. Though the gender gap varies a bit each time I teach this class, the average woman usually reports spending about three times as much as the average man. When I asked my students whether they thought it was fair that women, who already tend to make less money than men, should have to spend so much more on their appearance, they balked. "Maybe," one young woman offered, "it evens out because men buy women drinks." That would have to be a lot of drinks.

At his editor's request, James Cave, a *Huffington Post* writer, spent a month "pretending to be his girlfriend"—which translated to buying all of the "bathroom things" she buys in a month. Cave was stunned at the wide array of beauty products his girlfriend used on a regular basis, especially since he generally considered her to be a low-maintenance type. Scrubs, lotions, under-eye dark-circle corrector, mascara, concealer, blush, makeup remover . . . it was a longer list than he had anticipated, and the costs really added up. Not surprisingly, many of those who commented on the article noted that the majority of the products Cave's girlfriend bought were "not necessary." I understand that argument. There is no law requiring women to use antiwrinkle cream, and the need for mascara isn't the same as the need for sanitary supplies. But these commenters are missing a key point. For many women, these products *feel* necessary, and that feeling is a direct result of the culture in which we live. The nonprofit Renfrew Center Foundation found in a survey of around 1,300 women that nearly half feel unattractive or insecure when they go without makeup. A beauty-sick culture leaves women feeling like they need these (or other) products just to look minimally acceptable to the outside world. Walking out the door without mascara might feel like walking out the door without wearing pants.

Makeup may not literally be required of women (although for a handful of jobs, women are actually obligated to wear makeup), but if the other women you see are all wearing makeup, and every woman you see in every advertisement or television program or movie is wearing makeup, it's a bit facile to simply scoff and say, "No one is *making* you buy those products." Think back to the point Ana made in chapter 4. It's really hard to tell which products you actually want to buy and use versus which products you feel you need to buy.

Opting out isn't as easy as those commenters would have us believe. Just ask any woman who has gone without makeup only to hear, "Are you okay? You look really tired." Or ask any woman who has laughed when she hears references to a female celebrity's "natural beauty," knowing that natural look probably required at least an hour of effort by a professional makeup artist and specialized products most of us have never even heard of. Amy Schumer recently starred in a video that hilariously pointed out these ironies. In the sketch, she's approached by a boy band singing a song called "Girl, You Don't Need Makeup" to her. They go on about how she is perfect just the way she wakes up. At their encouragement,

Schumer proceeds to scrub her makeup off her face, but after she does so, the guys appear shocked and frantically recant their previous claims. They finish the song with "Don't take off your makeup. Wear it when you sleep and swim."

Given the pressure women feel to "keep up appearances," it should come as no surprise that huge amounts of money pour out of women's bank accounts and flow directly to the beauty industry. Women are responsible for 85 percent of the spending on beauty products. Market research reports estimate that cosmetics brought in over $60 billion in revenue in the United States in 2015, with over $8 billion on makeup alone. Whereas over half of men surveyed by YouGov reported using *no* products when getting ready in the morning, cosmetics company Stowaway recently reported that the average woman owns forty different cosmetic products.

Personal finance site Mint.com estimates that the average woman will spend $15,000 in her lifetime on makeup. One recent report by the YWCA made it crystal clear that money spent on beauty is money that could have been spent somewhere else.[1] According to their math, if you saved $100 a month that you would normally spend on beauty, after five years, you'd have enough for one year of in-state tuition at a public college or university.

We all need to buy clothes, but there's certainly a point at which clothing purchases qualify as beauty expenses rather than true necessities. When Jess started her first full-time consulting job, one of her coworkers ended up performing a fashion intervention of sorts with her. As Jess explained it, "She had to help me step up my game a little bit."

They would go shopping together and she would help Jess choose clothing and accessories. Once Jess had internalized those fashion standards and gotten used to shopping in more expensive stores, she continued that shopping on her own. She and that friend would call each other sometimes, saying, "Oh my god, I had an accident at Nordstrom!" *Accident* was code for spending way too much money at a given store. It was the job of the non-shopper to convince the shopper that the accident was acceptable because she *needed* those clothes for work.

Given how many women diet in attempts to change their appearance, it's fair to consider at least some of the money spent on diet plans as beauty money too. The vast majority of dieters are women, and the diet industry is more than willing to take their hard-earned dollars (approximately $20 billion of them per year[2]) in exchange for a drop of hope but little evidence of long-term effectiveness.

Cosmetic surgery and related procedures are another key area of beauty expense—remember that over 90 percent of these procedures are performed on women. Cosmetic surgeries (and less invasive procedures such as injections of Botox and fillers) are increasing in popularity so quickly that it's hard to keep up with the numbers. According to the American Society for Aesthetic Plastic Surgery, since 1997, the number of cosmetic surgeries obtained by women in the United States has increased 538 percent. In 2015, well over $1 billion was spent on breast augmentation alone.

The last time I made a routine visit to a dermatologist's office, I saw ads for all of the following beauty procedures: Botox, skin fillers, noninvasive cool sculpting fat reduction, radio-frequency wrinkle reduction, laser skin rejuvenation, liposuction, and upper and lower eye rejuvenation. Despite the fact that I was in the middle of writing this book, I still became aware of a growing list of facial flaws I hadn't even considered before entering the office. This type of marketing has compelling effects. When beauty feels like such a source of power for women, of course many women are willing to pay whatever they can to grab a piece of it.

Women Around the World Pay the Costs of Beauty

The United States often gets a reputation for being a hotbed of beauty pressures, as though we're the vainest of them all. This reputation is consistent with the strain of American individualism that emphasizes continuous self-improvement and the notion that there is always a better version of us hiding inside, just waiting to get out. Oprah Winfrey, one of the most influential and successful women in the world, recently took some well-deserved heat for a claim she made supporting this notion. In an ad for Weight Watchers, she explained: "Inside every overweight woman is a woman she knows she can be." As though an overweight woman could never be enough, just the way she is.

For women, the "better" version of themselves waiting to get out is almost always a thinner, more beautiful version—and that version doesn't come free. A nonstop barrage of marketing tells us that all we need to do to release that more beautiful version of us is spend money. The right mascara could change your life, we're told. The right wrinkle cream could stop time. A ten-pound weight loss could fundamentally change the nature of your romantic relationships and erase every neurosis.

Never forget that companies peddle beauty sickness in order to increase revenues; they have a vested interest in keeping you dissatisfied with your appearance. Their profit is directly tied to their ability to make you believe that their products will move you closer to the beauty ideal.

Beyond American-style individualism, there's another reason the United States is a somewhat unique contributor to women's beauty woes. U.S. marketers and the U.S. entertainment industry have played a major role in exporting a highly westernized, white beauty ideal to the rest of the world. From skin-lightening creams (projected to reach $23 billion in revenue by 2020) to double eyelid surgeries (the most common plastic surgery procedure in several East Asian countries), women around the world are paying to emulate the beauty ideal we sell, and that ideal doesn't come cheap.

As I gathered research for this book, I wanted to make sure I heard from women with an international perspective on beauty sickness. I put the word out via social media, and was contacted by Jaimie*. Jaimie has a particularly interesting vantage point on cultural beauty pressures, having lived in both the United States and South Korea. She was born in Alabama when her parents were graduate students, but moved backed to Korea when she was in elementary school. For college, Jaimie returned to the United States and stayed to complete a graduate program in graphic design. I caught up with her just as she was finishing this degree and packing to return home to Korea. We sat among half-filled boxes on a sweaty summer day.

When Jaimie is in Korea, she feels not quite Korean enough. When she's in the States, she doesn't really feel like an American. Even her chosen pseudonym captures this in-between-ness. Jaimie is the name she gives baristas at Starbucks in lieu of her Korean name, which Americans tend to mangle. This straddling of two identities has left Jaimie sensitive to beauty pressures from both of these very different cultures.

Jaimie first started thinking about beauty when her little sister was born. At six years old, Jaimie didn't really understand why, but her grandmother was seriously concerned about Jaimie's sister's appearance, calling her "not a pretty baby." A series of relatives made comparisons between Jaimie and her new baby sister. They indicated that Jaimie had been a beautiful baby, mostly because

Jaimie appeared to many as "half American" (which really meant half white), even though her parents were both Korean. Jaimie's baby sister was described as "foreign-looking." Although young Jaimie knew she was being complimented, she remembers these comparisons as triggering her first experiences of being self-conscious about how she looked.

I asked Jaimie about her sense that people thought she was a beautiful baby because she looked half American. "What features do you think they were looking at?"

Jaimie knows exactly what those comments were about. "I think mostly it was my eyes," she said. "They're still Asian eyes, but I have bigger eyes and I have double eyelid. I was born with that." Jaimie's talking about the crease that creates the folded-over look of some eyelids. Many East Asians don't have this crease, but it's considered highly desirable. There are eyelid tapes you can buy for a temporary effect. Many turn to surgery to permanently create this crease.

"You didn't have surgery?" I asked Jaimie.

"No." Jaimie smiles. "No, didn't have surgery. I joke with my friends that I saved a few hundred dollars being born this way."

Jaimie's little sister doesn't have double eyelids, but she wishes she did. Jaimie explains that sometimes her sister gets excited in the morning, because if she's really tired, her eyes will puff up in such a way that they look like they have double eyelids for a short period of time.

I asked Jaimie why she thought her grandmother was so worried about how she and her sister looked. From Jaimie's perspective, that worry was inherently gendered.

Jaimie explained, "She's a very traditional lady. She's very male-centered. So she's all about having sons. She was very upset that I was not a boy. When she found out I was not a boy, she started to care about my marriage."

"She was worried about what kind of man you would attract in marriage?" I asked.

"Yeah," Jaimie continued. "And the best way to attract a man is to be beautiful, I guess. So she was very worried about how I looked and how my sister looked."

Jaimie is matter-of-fact about this. It's not unusual in her world to hear that a woman's worth is defined only by the type of husband she can procure, and that looking a certain way is the clearest means of procuring the best husband. This is one framework through which we can understand some of women's beauty spending. If you feel that winning a good man is your only route to success in life, and if you believe that men are primarily won over by looks, the money spent on beauty feels like a solid investment. Jaimie told me that in Korea, there's often tension between the genders because "Women only treat men as wallets and men only treat women as trophies." In other words, she explained, lots of women feel they can trade their beauty for things, so lots of men feel it's open season to evaluate and comment on women's beauty.

Jaimie told me that lots of Korean students go on casual blind dates. "I have a lot of guy friends," she explained. "And they would share photos of their blind dates and grade them by how they look. And then they would make comments like 'I mean, if she looks this bad I don't even want to pay for her dessert or coffee. She's not even worth it.'"

Jaimie traces some of the gender difference in worries about appearance to a broader cultural phenomenon. "I think that the competition is a lot more fierce in Korea," she explained.

"Competition for what?" I asked.

"Everything," Jaimie answered, leaning forward for emphasis. "Korea is much smaller than America, and I think that's why everyone knows everyone's business and everyone wants to know whose son and whose daughter goes to which university. Since you're in elementary school, you keep being tested with all the exams. For Korean universities, there's a rank. A straight linear rank. So everyone wants to go to Seoul University, and it's no different for looks. Everyone wants to be the most beautiful."

To make things more complicated, Jaimie points out that before plastic surgery became so common, beauty was something you just had to be born with. "But now," she says, "everyone can fix it, everyone can be beautiful. I think that's why everyone is getting plastic surgery."

It's a surprise to many that the United States does not perform the highest number of plastic surgeries per capita. Instead, in most recent years, South Korea has led the pack. When Jaimie is in Korea, she's overwhelmed by the emphasis on plastic surgery. It's the first thing she notices when she goes home. Advertisements on buses, on bus stops, on the radio, on billboards. Everywhere you look, you see ads for cosmetic surgery. Jaimie says it's often the case that one bus or bus stop will be covered with nothing but ads for plastic surgery, with before and after photos taking center stage. Given all this emphasis, it's not surprising that somewhere between a fifth and a third of women in Seoul have had plastic surgery. Jaimie isn't one of those women. Yet.

Jaimie noted that attitudes toward plastic surgery among her fellow Koreans have substantially changed since she was young. When she was a child, hearing that a woman had plastic surgery would result in hushed voices asking, "Oh, why did she get it? Is she that unsatisfied with herself?" Now women openly discuss their surgeries. When Jaimie came to college in the States, she joined a cohort of around sixty other Korean international students. Jaimie was one of only four women in the group who *hadn't* had double eyelid surgery. Because so many people are getting this surgery, it gets cheaper and cheaper, Jaimie explained. You can get it done for about a thousand dollars. Still, a thousand dollars is a lot of money for most women.

But, as we know, Jaimie already has double eyelids—she was born that way. I asked her if she hadn't been born that way, would she have considered double eyelid surgery? Jaimie wavers a bit. She's not sure. But then she clarifies that it's not because she wouldn't have plastic surgery in general. "The plastic surgery I would consider would be more related to body than face," she explained. "Because I'm just not satisfied with my body. I would, like, definitely consider liposuction. And maybe boobs."

"You'd like larger breasts?" I ask.

"Yeah," Jaimie responds, "bigger boobs." But then she makes it clear that her breasts would just be the beginning. "I mean, I can go on and on. I'm not happy with my body. I know I have huge thighs compared to my upper body. I never wear skirts or shorts in the summer because I'm too conscious about my legs, which is pretty, like, pretty terrible. Sometimes it gets me really down. Especially during summer."

Jaimie laughs as she says this, a sad, resigned type of laugh. She calls the emphasis on women's thinness in Korea "obnoxious," but still feels bound by this standard. She casually tossed out that she had "starved herself" for a period of time during college, going as long as she could possibly go without eating anything, then, if she felt herself about to pass out, eating a yogurt. She lost weight, but got very sick and decided not to keep up with the regimen. She gained all the lost weight back, which is typical for most dieters, and now finds it even harder to maintain her set weight than she did before she experimented with starvation.

Jaimie still thinks she's fat, even though her mom, who is a nutritionist, tells Jaimie that she is perfectly healthy the way she is.

When Jaimie shared with her mom how bad she felt about her body shape, her mom was flummoxed. "I didn't raise you this way! How did you end up like this?" she asked. Jaimie explained to me that though she was raised by her mom, she was also raised by K-pop (i.e., Korean pop music). K-pop is where she learned what beauty looks like for women in her culture. Whether that beauty is healthy feels irrelevant. Jaimie explained, "I mean, my mom says, like, 'You shouldn't listen to all that!' But still, it's hard to ignore those things when you're actually living the reality."

Jaimie told me about a K-pop group called Girls' Generation. There are nine women in the group, and they've been at the top of the charts since Jaimie was in high school. Jaimie encouraged me to google the group so I could get a sense of what Korean women are expected to look like, so I did. They look like mannequins, nearly identical in shape. They are all extremely thin and long-legged. I found arguments on K-pop websites about whether *any* members of the group had not had significant facial surgery.

I can't get the image out of my head of Jaimie looking at a poster of that pop group and thinking to herself, "I'll just stop eating. I'll get liposuction. I'll get a boob job. I'll get my nose narrowed. I'll do whatever is takes." Jaimie told me it's on her bucket list to get a body she can be really happy with. Even if she has to pay for it.

Ticktock

I was raised by an Energizer Bunny of a single mother. She was a dedicated, award-winning elementary school teacher, steadfastly working with some of the least privileged students in her school system. My mom generally drove my brother and me to school each morning, dropping us off early, then heading to her school. When

I began thinking about the price of beauty, I reflected on different women I know and how the time cost of beauty pressures shows itself in their lives. When it came to my mom, the most obvious memory was of our morning drives to school.

My mom would bring her makeup bag in the car and put on mascara, lipstick, and other essentials at stoplights or occasionally even when the car was moving. Once in a while, she'd paint her nails while driving, hands splayed across the steering wheel. She never crashed the car as a result of her makeup maneuvers, though looking back, I feel we were lucky. When I asked my mom if it would hurt her feelings if I included this memory in a book, she joked, "Nah, I think it's too late for anyone to take you away because I was a dangerous parent. The statute of limitations has expired."

Young me definitely asked my mom, as we barreled down the road and she applied eyeliner: "Why do you have to put on makeup while you're driving?" But the more interesting question, one I didn't know to ask at the time, would have been "Why do you have to put on makeup?"

A substantial proportion of the adult women I interviewed for this book also casually reported applying makeup while commuting to work. The behavior is hardly unique. But it seems emblematic of how much pressure women feel to dedicate time to beauty work every day. These time costs of beauty may be even more problematic than the financial costs. Some women have an abundance of money. No purchase of wrinkle cream will detract from their ability to meet their basic needs or the needs of their loved ones. But I don't know one woman who doesn't yearn for more hours in the day. We're all perennially pressed for time.

A University of Maryland time diary study of over 1,000 U.S. men and women found that men have significantly more "free" or leisure time each day than women do.[3] That free time totaled

164 more hours per year, which adds up to the equivalent of four weeks of paid vacation for a full-time worker. Women are also more likely than men to report that they "always feel rushed." It's hard to find good scientific research estimating the specific time cost of the pursuit of beauty, but a less formal survey conducted by *Today* found that the average U.S. woman spends fifty-five minutes per day getting ready. That adds up to two full weeks a year. A *Women's Health* survey covered that finding with what felt like disingenuous shock. Their answer to the problem of too much of women's time being taken up by beauty? Timesaving tips like swapping out a regular hand lotion for a hand salve that lasts longer.

A British marketing research survey claimed that, on average, women spend almost two years of their lives just applying makeup. Beware of this estimate, though—the sponsor of the survey was a company selling spray moisturizer, marketing their product with an emphasis on how much valuable time it could save busy women. You don't even have to rub it in! Just spray!

After calculating how much money they spend on beauty, I ask students in my "Psychology of Beauty" class to add up how much time per week they spend on their appearance. It's not uncommon to see a gap between men and women of two to five hours a week. "What are you doing with all that extra time?" I asked the men. Playing video games and sleeping were the most common responses. The women in the class thought longingly of that additional sleep.

Beauty Time in the Workplace

Jess, the IT director you met at the beginning of the chapter, reported explicit evidence of the additional time her women colleagues devoted to appearance. We generally think of women's beauty time as occurring privately, in front of the bathroom mirror (or, ahem, the rearview mirror). But in this case, Jess witnessed a very public gender difference in beauty time. When she was working for a start-up, there was a culture of having a cocktail hour at four p.m. every day. Jess told me that at three-thirty sharp, the women would all go to the bathroom to reapply their makeup. "So you have to bring makeup with you to work?" I asked. Most days I can hardly remember to pack my lunch.

Jess laughed at my shock. "You've gotta bring makeup, you've gotta bring whatever your hair product of choice is. Some of the women would, legit, bring curling irons. And they would go into the bathroom at work and curl their hair. So not only are you primping for this company thing that you should just be able to get up out of your desk and go to, you're also bringing all kinds of accessories. And you can't have a meeting in that half hour. You can't do anything that's actually productive in that half hour that you're spending to primp yourself."

I'm still trying to process her story. "What are the men doing during that half hour?" I ask.

"Work!" Jessica exclaims, making it clear how ridiculous she finds this gender gap. "They're doing work. There's no change to their normal day. So in a way, it kind of debilitates you from actually being productive and successful. Because now you're spending work time to look pretty."

The culture of this particular firm was likely somewhat unusual—most of us don't have daily happy hours. But you can see this same type of beauty time gap in almost any work setting. Consider an entirely different workplace as an example: broadcast news. Think of how long makeup takes for a woman news anchor versus a man. In that same context, imagine the extra working out that is required due to news anchors' wardrobes. A man's suit hides a lot more than a woman's skintight sleeveless sheath (apparently the uniform of national newswomen). One outfit comfortably allows for a fairly wide range of body sizes and shapes; the other does not. Or ask yourself when the last time was that you saw gray hair on a female news anchor versus a male news anchor. The time costs of these gender gaps are real and significant. As former news anchor Christine Craft said, "Female anchors can stay on the air as long as men. However, they are required to have two facelifts for every one their male counterparts get."

Harrah's casino was recently sued by a long-serving, well-liked bartender who was fired when she refused to comply with a new policy that required women to wear makeup. The guidelines, called "Personal Best" by Harrah's, required nothing more from men than keeping their hair short; their fingernails clean, short, and unpolished; and their face clean of makeup. Women working in the beverage department, on the other hand, had a long list of grooming requirements, all of which cost time and money. They had to apply powder, blush, mascara, and lip color. Their hair needed to be "teased, curled, or styled" and "worn down at all times." In any job where public-facing appearance is even a bit relevant, women will require more time and money to get ready for work compared to men.

Jess, who shared the happy hour story, is torn about what she'll be willing to pay for beauty as she ages. She told me about an old college roommate with whom she's still in contact. "She gets prettier every year," Jess said. "It doesn't make sense to me, her regimen, how it could be working this well. She is literally anti-aging."

"Did she sell her soul to the devil?" I laugh.

"Totally." Jess laughs too. "She did. The plastic surgery, the nose jobs, and the fake eyelashes and all those things that you can do to yourself. To me, any time I've actually seen it, it has been so overblown and overdone. Then I find it unattractive. But not with her."

"She's doing it well?"

"She's doing it well," Jess confirms. "And in, like, such a subtle way that it makes me think, 'Well, maybe that is something I should consider.' Which is strange, it goes totally against every principle, everything screaming in my body is, like, 'No! You'd don't do that!'"

"Why not?" I ask. "What principle tells you not to do that?"

Jess gets quieter, struggling to put her response into words. "Because you should be proud of who you are," she begins. "There should be no reason to be ashamed of aging. You're learning about yourself, your life is changing. You know, yes, I worked to lose baby weight after I gave birth, but it wasn't, like, 'Oh, I'm gonna get a tummy tuck, and I'm gonna get my stretch marks taken care of.' You know, I earned those stripes. You earn your gray hair. You earn your wrinkles. And I don't ever wanna be ashamed of that. But at the same time . . ."

Jess trails off, shrugs her shoulders. "You're leaving the door open?" I ask.

"Yeah," she says with an air of resignation. "I guess so."

Finding a Deliberate Balance

Given how much praise women receive when they put in efforts to meet the cultural beauty ideal, it's unsurprising that it's often hard to tell when beauty practices are freely chosen and when they're less freely chosen. Most of us have never known what it's like to live in a world where our appearance wasn't paramount. Who knows what types of choices we might make if things were different? What might matter more to us if we lived in a world where our looks weren't such a major source of social and economic currency? Rather than ridiculing women for spending "unnecessary" funds on beauty products, we're better off focusing on the factors that make women feel obligated to spend their time and money in a never-ending quest to like what they see in the mirror a little bit more.

In one course I teach, I show a clip from a documentary called *Mansome*. One portion of the film follows a thirtysomething man who's openly obsessed with his looks. He can't let go of the idea that it's possible to become a ten on the appearance scale, and that with more time, more products, and more medical treatments, he might make it there. He gets facials and manicures. He gets his eyebrows threaded. He employs a cornucopia of beauty products. He puts in hours at the gym. He looks in the mirror *a lot*. Students watching this portion of the film often find themselves stifling giggles at this man, even as it becomes clear that he's in distress. We need to ask ourselves why it seems funny or embarrassing when a man pays this high a price for beauty, but normal when a woman does it.

When I ask young women in my "Psychology of Beauty" class whether they would like some of their beauty money and time back, they invariably say yes. To be clear, some women enjoy makeup or expensive salon services. They wouldn't sacrifice these purchases and are happy to give up the time they require. But we don't have to engage in all-or-nothing thinking here.

Many women enjoy and feel empowered by some beauty practices. Other practices seem like less of a choice; they feel more culturally mandated. Some of these practices (or products) are low-hanging fruit. Knowing that time and money are precious commodities, consider grabbing some of that fruit and taking it back. Hold it in your hand and ask yourself, "What do I most want to do with this time and money?" Maybe you want that time to spend with your friends, family, or significant other. Maybe you want that money to go into a vacation savings account, or maybe you want to donate it to a charity you care about. Maybe you want to use it to enroll in a class to learn something new or to start a new hobby. The goal should be letting your deeply held values guide your life choices more than a set of cultural expectations you never asked to be mired in.

As long as we continue to see women as objects, we have no need to question the necessity or desirability of different beauty practices. As long as young girls learn that their looks are always being evaluated by others, they'll grow up to become women willing to pay any price for beauty.

Though teenagers are still a minority of plastic surgery patients, tens of thousands of cosmetic surgeries (including breast augmentations, nose jobs, and liposuction) are performed on girls eighteen and under in the United States each year. When Scott Westerfeld's book *Pretties* painted a picture of a post-apocalyptic world where everyone gets plastic surgery at age sixteen in order to meet a highly rigid beauty ideal, the idea didn't seem that outlandish. In one of my favorite parts of the book, the protagonist, Tally, tells of a dream she had about a princess. The princess, stuck in her tower, becomes fascinated by what she sees out her window. Westerfeld writes, "She started spending more time looking out the window than at her own reflection, as is often the case with troublesome girls." If we want to raise girls to be the kind of troublemakers who hope to get out in the world and change it for the better, we're going to have to build a culture that discourages them from becoming obsessed with their own reflections. The costs of all that beauty monitoring are numerous. The price is too high.

THREE

This Is How the Media Feeds Beauty Sickness

7

Malignant Mainstream Media

WHEN I WAS a sophomore in college, I worked as a cocktail waitress in a local bar. As far as jobs go, it was pretty fun, and a nice way to pay my rent—even if it did result in my arriving home at two in the morning, sticky with spilled booze and smelling like an old ashtray. The bartenders were all men, and they treated me like a little sister, even going so far as to pat me on the head or ruffle my hair on occasion. One evening when I showed up for work, I saw that someone had cut apart a calendar of bikini-clad models and papered the entire employee area behind the bar with these images, Scotch tape clinging to the old wooden paneling.

Maybe today that type of display would be destined to land on social media and provoke a sound shaming of its creator, but this was a different time period. I didn't say anything to anyone about the pictures. In part, that was because I didn't have particularly high expectations for the workplace climate in a crowded bar, but it was also because I didn't know what to do or say. I couldn't even have told you whether there was anything truly wrong with what had happened. I just knew that the pictures made me feel shamed and vulnerable in a way I couldn't really articulate. They made me look at those brotherly bartenders in a different light and wonder how those same bartenders were really looking at me. Every time I went behind the bar to refill my tray with drinks, I was smacked in the face with those images, and every time I walked back into the crowd, I felt a little worse.

I never talked about the pictures with any of the other servers (all women), but I imagine now that I couldn't have been the only one feeling uncomfortable. The pictures were gone the next night. I don't know who took them down or why. I wish it had been me, but I was years away from being bold enough to make that kind of move.

Soon after the calendar incident, I saw a film by Jean Kilbourne called *Still Killing Us Softly: Advertising's Image of Women*. I was transfixed. In the film, Kilbourne masterfully wove image after image into a compelling narrative about how media portrayals of women systematically change the way women feel about themselves. She gave voice to the unease so many women feel when thumbing through the pages of a fashion magazine or watching yet another poreless woman hawk mascara in a television ad. She helped me understand why I felt the way I did that night at the bar, and confirmed that I likely had company in feeling that way.

I loved that film. It's what first opened my eyes to the role media play in promoting beauty sickness. We take the movement to hold media accountable for granted now, it's so widespread, but this is actually a relatively recent phenomenon. In many ways, the early versions of that film (which is now in its fifth edition) are what brought criticisms of media images of women into the mainstream.

If you're looking for someone or something to blame for the fact that so many women feel awful when they look in the mirror, mainstream media images of women are the easiest target by far. From the fitness models in bikinis pitching weight-loss shakes to the airbrushed-to-perfection cosmetic ads selling eye shadow

that promises to change your life, they might as well be screaming, "I am designed to fill you with self-doubt!" Famed street artist
Banksy elegantly described the power of advertisements to shape
self-perceptions, suggesting advertisers make "flippant comments
from buses that imply you're not sexy enough" and are "on TV
making your girlfriend feel inadequate." But more important, he
noted that this bullying is particularly effective because advertisers "have access to the most sophisticated technology the world has
ever seen."[1] This is a serious amount of power concentrated in the
hands of people who for the most part are much more interested in
separating you from your money than in improving your life in any
meaningful way.

I get several emails a week from junior high, high school, and
college students around the country who are doing school projects
on media images of women and looking for input. They're not being
assigned this topic—it just feels so relevant and important to young
women today that they regularly seek it out when given the opportunity to explore something that matters to them.

When I started studying the relationships women have with
their bodies, media images were absolutely my number one bad
guy. The Internet was still a toddler when I began this work (I had
to carefully cut images out of magazines for my research, as there
were no Google images to use), but even without Instagram or
Facebook, you could barely turn your head without seeing an airbrushed, shiny-haired, super-thin model selling something. Of
course these images aren't limited to advertisements; they inhabit
television programs, movies, and other forms of media as well. The
term researchers use to capture these types of depictions is *idealized media images of women*.

Idealized media images of women are far from being the *only*
important target when it comes to our beauty-sick culture, but

their sheer ubiquity means we can't underestimate their impact. We also cannot pretend that what we see in the media doesn't shape our thoughts and behaviors. It might be tempting to think that your mind is locked behind some protective wall, safe from the influence of the media onslaught, but that's not how brains work. We are *all* affected by these images. Their influence is insidious, and there is no magic force field to keep it out.

Without the breathtaking power of mainstream media to flood our eyes and ears with images of a narrow range of female beauty, it's impossible to imagine a world where beauty standards for women could have become so distorted. It's unfathomable that the cultural obsession with women's body size could have grown this intense without advertising to push it along. Women's body dissatisfaction could not have become so widespread without having been fed by an unending stream of rail-thin models interspersed with ads for diet products and miracle fixes for "problem areas." There is nothing natural about the current beauty climate for women. This is not the way it has always been. And we could not have gotten here without significant media intervention. When it comes to their influence on beauty sickness, there are three serious problems with media images of women. First, they are unrealistic and unrepresentative, distorting our sense of what women in the world actually look like. Second, they are unfailingly paired with images of success, romance, and happiness—continuously reinforcing the notion that a specific type of beauty is the key to a good life. Last, and perhaps most important, the women in these images are frequently sexually objectified, reinforcing our tendency to see other women and ourselves as things. Each of these media trends has been linked to a host of negative psychological outcomes in girls and women.

What World Is This?

Let's start with the criticism of media images of women that most everybody already knows: The images are not realistic. Being unrealistic means several things in this context. First of all, most media images feature women who are statistical outliers in terms of their appearance: they're young and they're unusually tall, with clear skin, shiny hair, and appealing facial features. Even if you're in New York or Los Angeles, you will almost never see a woman who looks anything like the women we regularly see on billboards, in television commercials, and in magazines. It's not that these women don't exist—they do, they're just extremely rare. Once a colleague told me he didn't think the women he saw in the media looked that different from the women he saw in everyday life. I nearly fell out of my chair with laughter. That's how powerful our media environment is. It can literally cause us to forget what the real world looks like.

Take a careful look around you the next time you're walking in a crowded area. If you pay close attention, you'll see women of myriad body shapes and sizes, hair colors, facial features, and ages. It's easy to forget the actual landscape of women's appearances, because the range of what we see in media is so narrow.

Although having a beauty ideal for women is nothing new (every culture and historical period has one), the current beauty ideal is marked by how absurdly unattainable it is for nearly all women. And this distortion is most obvious when it comes to body size and shape: it's very thin and long-legged. Large breasts and a curvy butt are required as well—yet no detectable cellulite is allowed. More recently, the requirement to have visibly toned musculature has been added to this already preposterous standard. It's hard to point to any one actual woman who could simultaneously embody each of these requirements. Even celebrities who are admired for their beauty generally can't meet all these criteria. An actress praised for her curvy body will still be asked by producers to lose weight. A long-legged super-thin fashion model gets critiqued for being too bony. A celebrity fitness instructor is derided for not being curvy enough. Even the bodies closest to "winning" still can't win.

Women did not choose this insane ideal. It's simply what we've been exposed to from a very young age. It filters into our brains even if we don't want it there.

As the average woman's weight has increased over the past few decades, female beauty icons like models, Miss America contestants, and *Playboy* centerfolds have actually gotten *thinner*. The scene on television is no different: One analysis of central female characters in eighteen primetime sitcoms found that 76 percent of the actresses playing these characters were of below-average weight.[2] A different study of top network television shows (including over 1,000 major characters, 56 programs, and 275 episodes) found that 1 in 3 central women characters were underweight.[3] As a point of reference, less than 3 percent of U.S. adult women are underweight.

Magazine images follow the same pattern. Two researchers from Arizona State University once designed a study to test how viewing images of average-weight models would influence women. Struggling to find such images in mainstream beauty magazines, they were forced to turn to specific "plus size" sources to find pictures of models who were simply average-sized.[4]

Not surprisingly, the clearest evidence of this pattern comes from the modeling industry. Researchers from Saint Mary's University analyzed data from modeling agency websites and found that over 80 percent of the models were underweight or severely underweight, with around 20 percent meeting what many consider to be the weight criterion for anorexia nervosa.[5] This is just a smattering of the research that all points to the same general conclusion: In terms of women's body size, what we see in the media bears little resemblance to what we see in real life.

Media images of women are also overwhelmingly unrealistic because of the work that goes into creating them—work that remains largely invisible to the viewer. These types of images almost always involve teams of makeup artists, hairstylists, hidden clothing clips, and strategically placed fans. But the work doesn't end after the pristinely monitored photo shoot happens—the images are also airbrushed and graphically manipulated. In other words, we start with women who are statistical outliers, and then these women are transformed from highly unusual into fictional. On occasion, women in media images are created entirely from scratch by cobbling together carefully selected body parts, becoming avatars instead of models.

Thanks in part to media literacy campaigns and savvy social media users, it's become basic knowledge that nearly all media images of women are airbrushed. I don't know any woman today who would be surprised to learn about Photoshopping. We know it's happening. But that doesn't stop advertisers from becoming more and more absurd in their manipulations: editing a shot of a model to make her so thin that her pelvis becomes more narrow than her head; taking off part of an arm and forgetting to replace it; creating hollowed-out armpits with no creases of skin; enlarging eyes and lips until a woman's face looks like anime; shrinking the waist of an already tiny model until the body presented couldn't actually support human life. A recent, particularly disappointing trend involves hiring extraordinarily thin models, then airbrushing away their visible ribs and collarbones, giving the false and potentially dangerous impression that a seriously underweight body can still appear healthy.

That's the beauty ideal we're dealing with here. It's highly un-likely, if not physically impossible, that anything—hundreds of crunches, meticulous calorie counting, extensive makeup, or even plastic surgery—is going to turn you into that ideal. You must re-member that the media-promoted beauty ideal for women is un-attainable *by design*. Our eyes are drawn to these images precisely because we don't see them in real life, and advertisers know this. The vulnerability that results from seeing these perfected images is what drives much of women's consumer behavior. Advertisers need you to feel yourself falling short of the ideal, because then you'll consider buying a product in an attempt to move yourself closer to it. I regularly see women scoff at the claims in beauty ad-vertisements, saying, "That's stupid." But if they gaze at the model in the image long enough, the next words out of their mouths are often something like "I wonder if it works?" This is perhaps what's most upsetting about the beauty ideal we see in the media—not just how widely accepted it is, but how effective advertisers are at con-vincing us that the ideal is, in fact, attainable.

An analysis of over ten years of *Seventeen* magazine revealed a focus on the creation of "body problems" in the minds of girls, a laundry list of body areas that need to be managed and nightmare scenarios about bodies gone wrong.[6] These types of beauty maga-zines (and their supposedly health-focused counterparts) system-atically create two sets of body characteristics: one desirable and one undesirable. Long legs good. Short legs bad. Flat belly good. Belly rolls bad. What's most egregious is how many of these mag-azines also contain articles encouraging girls and women to "em-power" themselves by fighting against the very same body ideals promoted in their pages.

Another unrealistic aspect of media images of women concerns how these images utterly fail to represent the actual diversity of women present in the world today. *Symbolic annihilation* is the powerful term that captures the extent to which members of different groups are simply absent from the media. Heavier women are often symbolically annihilated, but so are women of color and older women. One recent analysis of popular women's magazines focused either on fashion and beauty (for example, *Elle* and *InStyle*) or health and fitness (for example, *Women's Health* and *Fitness Rx*) found that over 90 percent of the female models in the magazines were white and around 80 percent were under the age of thirty.[7] It hurts when you can't look at mainstream media images and find anyone who looks a bit like you. It can make you feel unwanted and invisible.

Sasha*, a twenty-five-year-old woman working in marketing, knows what it's like to be made to feel invisible and inadequate by the dominant culture's media images. Her story of beauty sickness began early. One of Sasha's birth parents was black; the other was biracial. She was adopted as an infant by two white parents. Sasha alternates between black and biracial when referring to herself, but as she put it, she "never gets mistaken for a white girl."

Sasha met me at an out-of-the-way little crepe restaurant when she got off work. Having recently been burned by a malfunctioning recorder that erased an entire interview, I set up three different recorders on our tiny table, in between our plates. Sasha laughed as I did this. She conducts interviews and focus groups for her job, so she knows the perils of misbehaving technology. With her encouragement, I activated the recorder on my cell phone for backup to my backups. We talked in between bites of food, our tiny table looking like an electronics display.

Sasha's siblings are white. The town she grew up in and the schools she attended were populated almost exclusively by white people. She was the only person of color in her grade for many years of her childhood. The lack of diversity in her everyday life undoubtedly shaped her self-perceptions. But what Sasha focuses on when she reflects on her childhood is not that her classmates were all white, but that almost all of the women in the media she consumed as a child and adolescent were white. She never saw herself reflected in those images.

Sasha's earliest memories of worrying about her appearance started around age five or six. She explained, "I have distinct memories of my mom, like, trying to figure out how to do my hair. The daily struggle of trying to figure out how to do my hair." I asked what young Sasha thought the source of this struggle was. Sasha shook her head and laughed as she explained, "I thought the problem was my hair, a hundred percent. I mean, yeah! I thought the problem was my hair—not that my parents didn't know how to do it. Later on in my life, when I went into middle school, I all of a sudden really, really wanted to have straight hair. I did everything I could. For years, my parents would pay, like, $300 for me to get extensions and cornrows, so my hair was down to here and then I could put it up." Sasha reaches around and taps the middle of her back to show me how long her braids were. "All I wanted to do was put it up in a normal ponytail. That's all I wanted."

Sasha frowns when she explains this part of her childhood, like she's still working through it and trying to understand what happened. I noticed she used the word "normal" to describe the type of ponytail she wanted. When I asked her what a normal ponytail would look like, it became clear that by normal she meant "white." Sasha didn't want it to be a "poof" on the back of her head. She remembers exactly how she felt back then, telling me, "Well, obviously, the way that girls with straight hair, or whatever, were doing their hair, it just didn't look the same on me. And I just remember feeling very uncomfortable once I realized that was the case. Like, once I realized, 'Oh, I'm different in this regard.'"

This hair-related angst was intensified when Sasha began taking ballet lessons. Sasha and her classmates were expected to wear their hair in buns. Sasha recalls, "I had to put so much gel in my hair to get it to be smooth all the way back. I had ballet teachers always picking on me because my hair wasn't as, like . . ." She motions with her hands, pressing them hard along her head as though smoothing back a ponytail.

"It wasn't flat enough?" I ask.

Sasha drums her hands on the table for emphasis. "Not flat enough! And I was very, like, 'Am I just not doing this right?' It just felt, like, especially in that environment, that at any minute somebody was going to realize that I wasn't a good, put-together dancer. You know?"

It wasn't just her hair. Sasha also had a sense that her body was too far from the dancer ideal and didn't look like her classmates' bodies. She explained, "When I started to develop, looking back, comparing myself to my friends, I was just a little bit heavier than them. So that, combined with the skin-color thing and the hair thing just made me feel, like, awkward. I just felt awkward. I know now that a lot of young girls feel awkward."

I agreed with her, "At that age, yeah. Definitely."

Young Sasha didn't realize other girls might have also felt awkward. She tensed, reliving those moments. "I felt awkward in my own skin, but to me, it felt unique. I felt like I was the only one."

Sasha had a *CosmoGirl* subscription when she was a tween. She learned from that magazine and others that normal hair was white hair. She also learned that makeup was for white women; on her skin, it never seemed to work quite as it was supposed to. I could hear her exasperation. "I would look in magazines and go to the store and try to buy the same stuff that the other girls were buying, and it didn't look the same on me. First of all, the tutorials and stuff in like, *CosmoGirl*, at the time it was all white girls, you know? Until recently, they didn't even sell a foundation or a concealer at CVS that was even my color. I think early on it was just not having any role models. I felt uncomfortable, because I felt like 'There's going to be nothing that looks good on me anyway.' It just made me feel embarrassed, you know?"

At one point, Sasha purchased a Bobbi Brown makeup book. It gave her hope. She thought, "This is it. Once I read this, I'm gonna know how to be a woman. How to look like a woman."

"Look like a woman in general, or look like a white woman?" I wondered.

Sasha sighed. "A white woman, probably. And I never figured it out, how to look like that. Obviously." Sasha told me that her perceived failure to live up to the beauty standards she had bought into left her feeling horrible and embarrassed.

Sometimes Sasha feels that other women out there are born with some type of superpower that makes them look "together" all the time. They don't seem to sweat. Or wrinkle their clothes. Their skin looks nice and smooth. Their hair is never out of place. The media images Sasha grew up on clearly bear a big portion of the blame for this feeling. Where else could someone get the idea that it's even possible to look "perfectly put-together"? When we see these perfected images of women over and over, the line between fantasy and reality blurs.

"Maybe I just haven't figured it out?" Sasha mused. "Sometimes I think there is like a whole secret out there, a secret about looking a certain way . . . just put-together and polished. Maybe looking white. But it doesn't have to be a white thing. Is there something that people haven't told me?"

"If you find out, let me know," I responded. Though Sasha's frustration is palpable, I can't help but laugh a bit, because I've certainly felt the same way at times. The unspoken beauty rules for women often confound me, and just like Jess, the IT director you met in chapter 6, I've routinely wished there were some type of uniform for women so I could spend less energy trying to figure out what the rules are in different situations.

Sasha laughed too. But I got the feeling she was serious, to some extent. A part of her still wants to look like those images in *Cosmo-Girl*. She wants to know their secret. She wants in on it.

The Message Sinks In

Sasha's longing to meet the dominant culture's beauty standards and her shame at failing to do so are not unusual. Scientists have known for decades that the type of imagery I've described is affecting women in important and disturbing ways. The most clear and well-studied impact is on women's body esteem. An analysis of twenty-five different published studies on the topic found that exposure to thin ideal media images leads to increased body dissatisfaction for women.[8] But body dissatisfaction isn't the only worrying outcome of these types of images. Research also links these images to increased feelings of depression and anger as well as decreased positive emotions and self-esteem—especially among adolescents, an already at-risk group.

Images of women on television may have an even bigger impact than static images. A University of Illinois study found that female college students' perceptions of the ideal body for women were associated with how much television they watched.[9] The more they watched, the more they found themselves wishing for bigger breasts and smaller waists and hips. Regardless of your current body size, you're more likely to *feel* overweight if you watch a lot of television, because almost every woman you see on television programs is very thin. Shows such as *Extreme Makeover* and *Bridalplasty* packed a particularly powerful body-shaming punch, as contestants on these programs "won" surgeries to move their bodies closer to the cultural ideal. A study of over 2,000 college women found that those who spent more time watching plastic surgery-focused "reality" programs had higher body dissatisfaction and engaged in more disordered eating.[10]

Some of the most persuasive research tackling media influence on girls' body image was conducted in Fiji, by Harvard Medical School professor Anne Becker.[11] Fiji was a particularly interesting culture in which to examine these issues because at the time the research began, large bodies were prized in Fiji. A plump build was seen both as aesthetically beautiful and reflective of a strong social network that cares for you and feeds you well. Becker wrote that during meals, family members would encourage abundant appetites, saying, "Eat, so you will become fat." Fijian girls would have found Sasha's worries about her body size confusing. If anything, they would have encouraged her to gain weight.

Throughout the early 1990s, there simply wasn't a cultural norm in Fiji that encouraged women to make their bodies thinner through diet or exercise. Anorexia and bulimia were unheard of. Becker studied adolescent girls in a province of Fiji that first gained widespread access to television in 1995. Most of the programs available were Western—for example, Australian soap operas and the original *Beverly Hills 90210* were popular. Those in the unofficial roles of the pretty/popular girls on these programs included actresses like Jennie Garth and Shannen Doherty. Those actresses weren't shockingly thin by U.S. standards, but their bodies at the time bore no resemblance to the Fijian body ideal. Becker and her colleagues collected baseline data from teenage girls within just a few weeks of television becoming available. Three years later, they returned and collected data from a new group of adolescent girls. At the baseline, no girls reported self-induced vomiting to lose weight. Three years later, 11 percent of girls did. After television had been around for three years, 74 percent of the Fijian adolescent girls surveyed thought they were too fat. Before television, it wouldn't have even made sense to ask these girls if they felt too fat.

How did watching TV so drastically change the way these girls from Fiji felt about their bodies? The same way it affects women in the United States or anywhere else with ready exposure to the thin body ideal. First, the more of this type of media you consume, the more likely you are to internalize the beauty ideal you see in the media. In other words, if you see that ideal everywhere you turn, you're more likely to buy into it—to make it your own. Buying into this beauty standard isn't just disheartening, it can be dangerous. Internalization of the thin body ideal shown in the media is one of the strongest predictors of the development of eating disorders—an association clearly demonstrated in the Fiji study.

A second route through which these types of images leave women feeling inadequate is called *social comparison*. Social comparison is a process we all engage in, just as Sasha compared her hair to the hair of models in magazines and her body to the bodies of her fellow dance students. Part of human nature is that we're motivated to know where we stand on a variety of different characteristics. Sometimes we have objective information available, like a score on a test. But for most of the characteristics that interest us, we don't have that type of data. Instead, we rely on informal comparisons with others to figure out where we stand. If you want to know how kind you are, you'd probably think of people you know and people you see in the media and get a sense of where your kindness falls relative to theirs. If you want to know how physically attractive you are, you do the same thing.

Because many women are already spending a good deal of time worrying about how they look, and because idealized media images of women are so readily available, social comparisons to media images are almost unavoidable. If you're standing at the checkout aisle at the grocery store and you're treated to a magazine cover featuring an impossibly beautiful celebrity, it's incredibly difficult to avoid being hit with a sense of how your looks stack up compared to hers.

Because the beauty ideal for women represented in media images is largely unattainable, social comparisons provide endless opportunities for women to feel they're falling short. Social comparisons tend to happen quickly and automatically, so they're very difficult to stop, even with conscious effort. When I was a graduate student, I was interested in getting a sense of how often women fell prey to these sorts of comparisons when they saw media images of women. I asked just over 200 college women to write down all the thoughts they had while looking at magazine advertisements from women's magazines. Some of the images just featured products, like an ad for lip gloss from Clinique with a tube of gloss spilling onto a minimalist white background. Others, like a Calvin Klein swimwear ad and a Clarins moisturizer ad, featured models. Social comparisons in response to images of models were extremely common—over 80 percent of the women made at least one. Just reading the list of comparisons these women wrote is heartbreaking.

- *I wish I had a perfect flat stomach like hers.*
- *Why can't I be that thin?*
- *I think I would be happy if my thighs were that thin.*
- *God, she's pretty. Why can't I be that pretty?*
- *She's every guy's dream. I wish that were me.*
- *I feel like a chunky elephant compared to this model.*

- *It makes me think about all of the aspects of my body that I hate. It makes me feel like I should change my appearance.*

Looking at these ads also led women to generate wish lists for their own appearance and a laundry list of their own perceived flaws.

- *I hate my hips, my chest is too big, and my stomach isn't flat enough.*
- *I wish I had a different eye color.*
- *I'm too fat.*
- *I hate having freckles.*
- *My skin needs more tone.*
- *My thighs are too big.*
- *I wish I actually had a crease in my eyelid.*
- *I have too much acne.*
- *I'm not pretty enough.*
- *My eyelashes are not long enough.*
- *My body is ugly.*

After working with a team of research assistants to code all of the thoughts the women in this study listed, we learned that those with the highest levels of body dissatisfaction made the most social comparisons. When you're already feeling vulnerable about how you look, your tendency is to constantly seek out more information about your appearance. This creates a vicious cycle of body hatred.

Those who struggle to feel good about their bodies also show what are called *attentional biases* in response to media images of women. Their eyes are drawn to the bodies of thin women, and when looking at pictures of themselves, they can't seem to tear their eyes away from the parts of their body they dislike the most. In one study my lab conducted, we first surveyed over 80 women about how satisfied they were with different areas of their body. Later, we asked them to look at images of thin female models from catalogues while seated at an eye tracker. Eye trackers use infrared light to provide extremely sensitive indicators of where your visual attention goes. When shown images of female models, the women in our study tended to immediately look at the area of the model's body that coincided with the area of their own body that they dislike. In other words, if a woman was dissatisfied with her own thighs, her eyes tended to immediately go to the thighs of the model she was shown. This pattern, which happened so quickly it was likely unconscious, undoubtedly contributes to the social comparison process.

When she was growing up, Sasha clearly felt the effects of beauty-related social comparisons in her life. After falling short of the beauty ideal one too many times, Sasha eventually concluded that *pretty* just wasn't something she was *allowed* to be. She decided to drop out of a beauty race she felt she could never really compete in. She explained, "I just kind of opted out. And I decided I would play kind of a different role within my friend dynamic." When I asked her what role she played instead, she knew the answer immediately.

"I was the sidekick," she said, shrugging her shoulders. "I know exactly what role I played, and I played it with so many different friends throughout, you know, middle school and high school. I was always the sidekick."

Sasha also knew exactly what it meant for her to be a sidekick. "For me," she explained, "it was, like, the girl who wasn't the one that people thought was necessarily pretty, but I was nice and I was funny. And I was always good at helping my friends get what they wanted. Whether it was a boyfriend or to ace their test. I was always good at helping that other person shine as the person they wanted to be."

In some ways, this sounds nice. Helping others, having a good sense of humor—these are admirable qualities. But if you listen carefully to what Sasha is saying, this doesn't feel so great after all. If she couldn't be the type of "pretty" the media was selling, she felt she would just be invisible instead. Sasha confirmed my concern, admitting, "It was to compensate. I became the secondary person in a lot of people's lives instead of shining in my own life because I felt like that was my role. I did want to be considered attractive. I just knew I wouldn't be. And I . . . I felt like secretly, that everybody kind of knew I was different and felt a little bad for me. I didn't like that pity."

Perfect Things

In 2009, *WAD*, a French fashion and culture magazine, celebrated their tenth anniversary by printing a cover image that looked like a realistic photo of a naked woman lying on her stomach, with full view of her back, legs, and butt. But this was a special naked woman: one made out of cake. The image featured a slice being cut out of her ass. Happy birthday, have a literal piece of ass! You can search "WAD cake woman cover" online if you'd like to see this master- piece. To make it even more absurd, the cake woman was so thin that her ribs were visible. She got to *be* a cake, but she didn't get to *eat* any cake.

Try a thought exercise to put all of this in perspective. Imagine a magazine cover featuring a nude male body made of cake, with a slice cut out of the derriere. It probably seems disgusting or hilar- ious in a way the actual picture does not. It's easier to turn a woman into a sexualized thing, in part because media images have been so effective at helping us become accustomed to this notion.

We need to ask why this particular type of artistic choice is made so often with women's bodies and almost never with men's bodies. You can't answer that question honestly without recognizing the gendered nature of objectification. It's not that we never think of men as objects; of course we do. But presenting men's bodies in blatantly objectified ways isn't the norm. Seeing women's bodies that way is.

Not every media image of a woman is sexually objectified, but enough are that it's easy to pick out examples. Most obviously, it's not uncommon for women in media images to be visually transformed into actual objects (pieces of furniture, beer bottles). On top of that, images frequently show only parts of a woman's body—a leg here, a breast there. A classic example is a Marc Jacobs advertisement featuring former Spice Girl Victoria Beckham's legs awkwardly sticking out of a shopping bag. All you see is a blackand-white shopping bag with stick-thin legs falling over the side like overflowing groceries. We know these limbs are Beckham's only because the advertiser tells us so.

Beyond creating unrealistic ideals and distorting our idea of what women actually look like, media images of women do another type of damage as well. They are one of the main sources of the sexual objectification of women, constantly conveying that women's bodies exist for others to evaluate and use at will. When we see women portrayed as objects in media imagery, it's a reminder of how often women are valued only for their bodies. The message of these images is clear: *You exist for being looked at.* More subtly, some of these images are shot in such a way that just by looking at them, we are forced to take on the perspective of someone doing the objectifying.

Those who worry about the objectification of women in media images were particularly outraged by a somewhat recent advertisement from Dolce & Gabbana. Anger over the image went viral.

It features a shirtless man pinning the arms of a barely dressed woman above her head, while three other men watch from the sidelines. The woman is clearly the focal point of the ad, but she barely looks present. She's staring off into the distance, almost seeming to dissociate. She doesn't seem to want to be there. All the while, she's stared at by men who look ready to take turns having sex with her. The tendency to portray women as not fully present, as gazing into space instead of seeing what's going on around them—even when what is going on around them seems critically important or dangerous—is all too common.[12]

Both Spain and Italy banned this advertisement when complaints surfaced that in addition to being demeaning to women, it appeared to portray a gang rape. The image was not banned from U.S. magazines, but even if you didn't see the two-page spread, you probably caught the ad somewhere online, since banning specific images has the tendency to draw more people's eyes to them.

Recently, the Tumblr page "The Headless Women of Hollywood" garnered a blast of attention on social media. The site compiles dozens of movie posters that prominently feature a woman's sexualized body, but not her head—yet another example of the objectification of women in media imagery. It's almost as though Hollywood forgot that women have faces.

The faceless trend for images of women is nothing new and is not limited to movie posters. When we see images of men, we are more likely to see their faces. And even when images show both men's and women's faces, they tend to show *more* of men's faces. This trend has been documented in a variety of magazine types around the world, in newspapers, in artwork, in prime-time television programs, and even in amateur drawings.[13] A recent analysis found that even on their own social media profiles, where women get to choose how to present themselves to the world, we see more of women's bodies and less of their faces.[14] For too many, seeing a woman's body tells you everything you need to know about her. A series of studies out of the University of California at Santa Cruz demonstrated that when we see images featuring more of women's faces (versus bodies), we assume the women in the images are more intelligent, more assertive, and more ambitious.[15] In other words, face-ism isn't a simple artistic choice. It has serious consequences for how we view women's character and potential.

Perceiving women as objects can be something that happens so quickly and automatically that we're not even aware of it. Cognitive psychologists have long known that we process images of things differently than we process images of humans. For humans, we tend to use something called *configural processing*. To recognize a specific face, we don't just need to see the eyes, the nose, and the mouth. We need to see how those different components exist in relation to each other. How far is the nose from the mouth? Where are the eyes relative to each other? As a result, images of humans tend to be harder to recognize when we flip them upside down, something called the *inversion effect*. On the other hand, objects (like a house, for example) are easily recognized even when they are inverted. In one creative study published in the journal *Psychological Science*, researchers showed participants sexualized images of both men and women, all of whom were shown in underwear or bathing suits.[16] Consistent with the inversion effect, participants had a hard time recognizing upside-down men. However, when it came to the sexualized images of women, participants did not show an inversion effect. These images were recognized just as easily whether they were upright or upside down. The sexualized women were *literally* processed in the same way that we process images of things.

The Consequences of Exposure to Objectified Images of Women

Social scientists make a number of different arguments about *why* women are so likely to be objectified in media images. Maybe we evolved to be more interested in women's appearance. Maybe marketers are just giving us what we want. Maybe it's blatant sexism. But no matter *why* you think women are often portrayed in sexually objectified ways, the consequences are disconcerting.

Even short bursts of objectifying media can have measurable impacts on the way we think about women. For example, in one study, researchers showed men one of two types of videos.[17] In the objectification condition, men watched movie scenes in which women were shown in sexualized or degrading ways (one scene was a striptease from the movie *Showgirls*). In the control condition, a different group of men watched cartoons from an animation festival. Afterward, both groups of men were told they would then participate in a separate study about decision-making. During this ostensible "decision-making study," the men read an article about a college woman who was date raped. The men who saw the objectifying movie clips were more likely to say that the woman who was raped probably enjoyed the sex, and that she "got what she wanted in the end." In a different lab study, researchers found that compared to men assigned to play Pac-Man or The Sims, those who spent twenty-five minutes playing a sexualized video game called Leisure Suit Larry later rated sexual harassment of women as more acceptable.[18] Men in the Leisure Suit condition were also quicker to identify words like *slut*, *whore*, and *bitch* in a word-identification task, suggesting that the objectifying imagery in the game primed them to think of women in these ways.

But objectifying images don't just increase our willingness to view women in a negative light; they can also decrease the extent to which we see women as capable, skilled human beings. In one study out of the University of South Florida,[19] college men and women were asked to write about either Sarah Palin (who was a vice presidential candidate at the time) or actress Angelina Jolie. Half of the research participants were told to "write your thoughts and feelings about this person's appearance." That was the objectification condition. The other half were simply asked to write about the celebrity they were assigned, with no specific instructions about appearance. Afterward, everyone rated the celebrity they wrote about on several dimensions. It didn't matter whether the subjects were writing about Palin or Jolie—if they focused on appearance, they rated her as a less competent human being. Consider the bind this puts women in. You live in a culture that demands that you focus on your own appearance and promises others will do the same. But this evidence suggests that the more people think about your appearance, the less likely they will be to view you as competent.

Additional evidence of the dangers of viewing women as things comes from a study conducted by psychologists at Princeton and Stanford. These researchers found that men who saw sexualized images of women were quicker to associate those images with first-person action verbs (*use, manage, grab*), but faster to associate nonsexualized images of women with third-person action verbs (*uses, manages, grabs*).[20] Think about it this way: You act upon *things* (I control her, I manage her, I use her), whereas humans can do their own acting (she controls, she manages, she grabs). The more sexualized a woman is, the more she appears to others like the object of an action instead of the actor herself. The more women seem like things, the more difficult we find it to imagine them doing things instead of having things done to them and the harder it becomes to see them as fully human.

* * *

SASHA NEVER SPECIFICALLY BROUGHT UP objectification during our interview. For her, the basic fact that the media images she
was consuming looked so little like her was overwhelming, leaving little need to interrogate the images much beyond that. But the
fact that Sasha was (and still is, to some extent) so concerned with
how others see her appearance suggests that the broader lessons
of these images have made an imprint on her thinking. After all,
you have to believe that your appearance is of utmost importance
to bother spending much time worrying about how others evaluate
it. That third-person perspective is part of the detritus these objectifying images of women leave behind. Sasha, like almost every
young woman I've talked with, has had to actively work at shedding
that internal mirror. She's had some successes, but the mirror is
definitely still there.

Sasha changed quite a bit in the years after she first got that
CosmoGirl subscription. By the time she was in her late teens, she
started trying to let those images go. She told me, "At this certain
point I realized, like, 'Oh, yeah, none of these people look like me.
That's why this is so much harder.' But I still read those magazines
for a while. At some point, I—I put them down. I put them away."
Sasha remembers this process as being a deliberate choice.

She reflected back on how she felt at the time. "I was just like,
this is not reality for me. This is not healthy for me. It may have
also been a bit of, 'This is never going to be me,' but also a sense of
'This isn't worthwhile, this isn't real, and I'm glad I'm realizing it
early on in life.' That ideal. I was never gonna—I was never gonna
actually approximate that norm. So, at some point, I was just like,
'This is BS.'"

I was so happy to hear Sasha call out the BS for what it was. "Yeah.
You have to get angry at some point," I told her. Sasha agreed, but
clarified that she's far from feeling like she's won yet.

"Yeah. So I say that. That it's BS." She paused, choosing her words carefully, "And yet I think there are still pieces of it that I'm frustrated by. You know? It was a lot of years I spent trying for that ideal."

How did someone as intelligent and interesting as Sasha come to believe that looking like a model in a magazine would change her life so drastically? You don't have to look far to find the answer to that question. Media images of women almost never exist in isolation. They're paired with the products and lifestyles we associate with living the good life. If you see images of two things paired together often enough, it's basic classical conditioning that you'll begin to associate one idea with the other. In addition to pairing the female beauty ideal with money, success, admiration, and love, mainstream media outlets regularly pair violations of this ideal with negative life outcomes or even straightforward ridicule. For example, analyses of over 1,000 characters on ten popular prime-time television programs revealed that thinner women are substantially more likely to be shown in romantic relationships, whereas heavier women are much more likely to be the targets of jokes at their own expense.[21] If we consume enough of this type of media, we become unable to separate actual success in life from how we feel about the image in the mirror.

Sasha decided it wasn't her fate to turn heads when she walked down the street. But knowing that, in a way, was freeing. It allowed her to think more about her values and beliefs. She declared herself a feminist as soon as she was "old enough to know what the word meant." And because she didn't have the privilege of white skin or long straight shiny hair, she became more sensitive to issues of privilege in general. She told me she was grateful for having always felt a bit like an outsider, "because it forced me to go into spaces where everybody was a little bit different in whatever way they were. That's how I started to pick up some of the things that I hold dear to me today and some of the things that are important to me today."

When I followed up by asking Sasha what's important to her today, she seemed surprised. This is one of those questions we just don't ask young people enough. She reflected for a moment, then responded, "I've always held being your true self to be very important. I deeply value the person within the exterior."

Sasha told me she's come a long way, because she can wear her hair down now and feel comfortable with it. She even took it down to show me, casually running her fingers through it and turning around in her chair so I could see the strands touching her neck. When we finished our interview, I asked Sasha if she wanted to choose her own pseudonym for this book, explaining that some people have a secret name they always wish they had. Choosing a pseudonym can be a fun way to claim that name for yourself.

"Do you have one?" I queried.

Sasha looked down. "When I was younger," she told me, "it was Kimberly. There was this girl that was a few years older than me, who I admired. Her name was Kimberly. She had bouncy blond hair. It's a white-girl name."

"Do you still feel like a Kimberly?"

"No," Sasha assured me, "I know I'm not a Kimberly. But I wanted to be a Kimberly for a long time."

That statement hung in the air for a moment while Sasha played with the food on her plate. But then she looked up, smiled, and nodded. As I write this, I have an image in my mind of Sasha presiding over a bonfire of those magazines that so bullied and misinformed her when she was a child. I see her finally breaking their grip on her self-image. I know a lot of women who would like to add to that metaphorical bonfire. The type of media imagery we talked about in this chapter is powerful, but it comforts me to know that women's desire to break free from the ideal these images spread is also powerful.

8

(Anti)social Media and Online Obsessions

WHEN I LOOK around at the selfie-saturated, social-media-driven world we live in today, I thank my lucky stars I was born when I was. The Internet didn't really emerge until I was in college, and even then it was sparsely populated and not particularly interactive. But more important, digital photography wasn't widespread until I was well into adulthood. When I was growing up, taking a picture of someone was much rarer. It took effort, time, and expense to get film developed. With the exception of professional photographers, most people carried cameras around only if they were on vacation. Pre-smartphone, we lived freer lives in some critical ways. Back then, you got fixed up for school picture day. Today, for many girls and women, every day feels like picture day. We have social media to thank for that feeling.

It's true that the cultural forces that make so many women feel like they're on display all the time were active well before social media and digital photography. It's not like street harassment wasn't happening, or media images of women were substantially more realistic or less objectified a couple of decades ago. Particularly for young women, even pre-social-media, everyday life was still full of reminders that a woman's primary form of cultural currency is her appearance. On more than one occasion back when I was in college, I walked by a fraternity house on campus and saw men on the front porch holding up signs with numbers on them, scoring the women who walked past. That was pretty terrible. But how much worse would it have been if they were wielding cell-phone cameras and uploading photos and ratings to some Reddit thread or Instagram account?

One of the biggest problems with a beauty-sick culture is the way it leaves women feeling like they have to be on top of every detail of how they look at all times. Knowing that there's always a camera ready to capture those details only ups the ante. Maria, a twenty-five-year-old woman you'll meet later in this chapter, described this climate as "almost like a paparazzi situation." The use of the word *paparazzi* is telling. Not only do women feel pressured to look like models and celebrities, now they can even experience a type of anxiety previously limited to celebrities.

One of the most fundamental ways that online images influence women is through activating those social comparison processes we talked about in chapter 7. Your peers will always feel like a more relevant standard of comparison than celebrities or models, one that's harder to dismiss. On top of that, images of our peers often feel more realistic than they actually are. It's easy to forget that many of these images have also been carefully posed, selected from dozens (or hundreds) of shots, Photoshopped, and filtered. The photos many women post today are created using the same tricks professional photographers and models have used for decades. We end up comparing our actual selves and real lives to the make-believe selves and pretend lives of others. The pressure to compare, to evaluate our lives in terms of where we think we stand relative to others, is difficult to escape.

* * *

MARIA IS A TWENTY-FIVE-YEAR-OLD WHITE woman with long dark hair, twinkly brown eyes, and a delightful grin. After spending most of her life in the Midwest, Maria recently moved to New York City to pursue a career in journalism. She's a graduate of Northwestern, where I teach, but I never met her during her time as a student. Though Maria's tale of beauty sickness begins before she became tangled in a web of social media images, today, Maria coruns a website dedicated to revealing the discrepancy between what we display in social media and what's really going on in our lives.

I spoke to Maria on the phone for over an hour, while I paced slow circles around my apartment. I might have guessed Maria was a writer based on our conversation. She didn't just answer my questions; she told me stories. Maria wove together pieces of her childhood, adolescence, and young adulthood into a complex and thoughtful tale of beauty sickness. She's done a lot of self-reflection recently, trying to figure out how and why she became the person she is today. Much of Maria's self-concept has been based on how she stacks up compared to others, a trend that emerged even before social media began playing a meaningful role in her life.

Maria was around eight years old when she started worrying about how she looked compared to other girls. She was in a dance class and some representatives came in from a company that sold clothes for dancers. On the hunt for catalogue models, they watched the girls dance for a while and then made their choices. Maria was not selected. She remembers that moment as one of the first times she realized that there was a metric being used to measure prettiness. She thought to herself, "Okay. Not as pretty as these other girls they wanted." It stung. She considered why she wasn't chosen. Maybe it was because she had such a big, toothy smile? The other girls' teeth were straighter, she concluded.

Maria told me, with audible compassion and gratitude, that her parents never emphasized appearance. Looks weren't a priority in her family. Instead, they focused on things like academic accomplishments. But as a child, Maria was nonetheless hyperaware of how she appeared to others. She was also highly driven, learning the pain of perfectionism at a young age. Maria wanted to be the best at everything she did. So once it became clear to her that others were evaluating her appearance, she thought, "Whoa. That's one more thing on this list of things I need to be good at." Being pretty became just another domain in which she felt the pressure to excel.

In other areas of her life, Maria felt she had a good deal of control over her performance. If she wanted higher grades, she could study more. If she wanted to improve at soccer, she could practice harder. Trying to excel in the "pretty" domain was much more frustrating. It seemed unfair to Maria that there was only so much she could do to improve her physical appearance. She reflects back, "I just remember that period of time when I felt so ugly, and yeah, I felt like it was unfair or something. I was, like, 'I can't improve this! Everything else about myself that I look at and want to change, I can change. And this, I can't.'"

On top of that, there was the frustration of not knowing exactly what the standards were for the implicit contest for prettiness or who got to set those standards. Maria told me, "I remember at different times when I was in elementary school, people would get in trouble for creating rankings of who were the prettiest girls in the class or whatever. And I remember wanting to be one of the people on the list. That's a vivid memory."

I asked Maria if she was on the list. She thought about it for a moment. "I think later on I was. There were different iterations. When we were older, like, junior high, I think I was. But, I remember, I think my general feeling about it always was 'I'm probably not one of the top couple, but maybe I'm pretty good.' I remember thinking, 'Okay. I'm not Molly, I'm not Brooke, but maybe I'm decent.' I just remember sort of getting a sense of how I stacked up."

"Were you okay with that?" I asked her, the incredulity in my voice obvious.

"At the time, I remember thinking it was silly. I didn't think it was good. I definitely thought, 'This is dumb and horrible that this has happened,' but I remember being fine with how I was evaluated." Maria speaks in a matter-of-fact way, as though the memory were stripped of emotion.

The practice of boys openly ranking the prettiness of their female classmates seems more than just silly to me. The fact that it happened at all feels more important than whether Maria made the list. "These random guys were in charge of deciding whether you met the criteria?" I pushed.

Maria doesn't seem to find the practice as offensive as I do. "Right, totally, they were," she confirmed. "But if those were the criteria, I wanted to be the best, you know? Did I think them making that ranking is dumb? Yes, I did. But since it's made, I want to be in the top five. You know what I mean? It's, like, if that standard exists, I better be the best at it."

Maria knows this kind of thinking is only intensified today, thanks to social media. She compared those elementary school rankings to the number of Twitter followers you have or the number of likes you get on a photo. "You know, do I think those metrics are dumb on some level? Yes. But since everyone else can see the number, do I want to have the most? Yes." It feels like basic logic to her.

How Much and What Are We Consuming on Social Media?

In 2015, a Pew Center research study found that 92 percent of U.S. teenagers spend time online daily, with the vast majority using more than one social network site. Among those between eighteen and thirty-four years old, one in five report spending *six* or more hours per day on social networks. Though both men and women can be heavy users of social media, women spend more time on social media than men do.

Television, movies, and magazines have staked their claim as enforcers of beauty sickness, but they're being nudged aside by their powerful social media counterparts. Social media packs a one-two punch of imagery and interpersonal interactions into one medium. Exposure to Photoshopped images of the female beauty ideal is bad enough. But think about how the impact of that picture is magnified when it is shared by one of your own friends, or when it's accompanied by commentary detailing every aspect of that woman's appearance. To some extent, it doesn't matter whether those comments are positive or negative. Regardless of whether someone writes, "So hot! So beautiful!" or "Fat and ugly," the message we hear is that commenting on a woman's appearance is important, or at the very least entertaining. And either way, it's definitely fair game. When you gaze at images of women all day and read all those comments focused on women's appearance, how can you *not* grow to believe that your appearance is under the same scrutiny?

Just as magazines and television programs tend to show only one beauty ideal, so too do online media. Sexualized images of women are rampant online, as are ads for weight-loss programs and plastic surgery. An analysis of twenty websites ranked as highly popular with U.S. teens found that the most common ads on the sites were for cosmetics and beauty products.[1] Nearly 30 percent of these teen-targeted ads were coded as emphasizing appearance, and 6 percent of the advertisements were for weight-loss products (for example, an ad titled "Tips for Losing Belly Fat"), even though these sites ostensibly were designed for young girls. In social media feeds, we have little control over the ads that show up—and advertisers are definitely targeting women with appearance-focused strategies. On my personal Facebook page, I flagged an offensive diet ad (featuring a drawing of an emaciated woman in a bikini and dubious claims about magic weight-loss pills) at least a dozen times before it stopped showing up in my feed.

In addition to widely used social media platforms, deliberate communities focused on weight loss and beauty proliferate online, offering more and more ways to spend time thinking about how you look or how to change the shape of your body. Two of the most flagrant of these types of communities are fitspiration (fitspo) and thinspiration (thinspo) sites. Thinspo sites tend to be populated with a predictable series of emaciated-looking models and inspirational quotes aimed at promoting starvation diets (e.g., "Every time you say 'no thank you' to food, you say 'yes please' to thinness!"). Fitspiration sites are ostensibly focused on health and strength, with pictures of six-pack abs and claims that the body of a fitness model can be yours if you'd just work harder at it (e.g., "Squat. That ass isn't going to get round by itself!"). A 2016 analysis of fitspiration and thinspiration websites found that there wasn't much to differentiate them. Fitspiration is generally just the poorly disguised cousin of thinspiration. Both seem to shame more than they inspire. Both types of sites contain a similar number of sexually objectified poses and messages stigmatizing weight, linking food with guilt, and encouraging harmful eating habits.[2] A 2016 study out of Australia found that one-fifth of women who regularly post fitspiration images were at significant risk for eating disorders. There's nothing wrong with wanting some fitness-focused inspiration, but if fitspiration was really about fitness, we'd see a much wider variety of body shapes in these images. My women students tell me their feeds are regularly cluttered with fitspo images. At times, they invite the images in by following popular fitness experts, trainers, or celebrities who post little else. Yet they seem to resent the images even as they sometimes can't look away.

Just past thinspo and fitspo sites on the beauty-sick continuum are pro-eating-disorder websites and discussion forums. These sites (often referred to as pro-ana or pro-mia) essentially promote eating disorders as a lifestyle choice and provide advice on maintaining and increasing dangerous eating habits. They also tend to be filled with images that could easily be found on thinspo sites. A study of over 700 adolescents in Belgium found that by ninth grade, 16 percent had visited a pro-ana website.[3] Visiting these websites tends to leave girls and women feeling fatter, uglier, sadder, and more anxious.

Think about it this way: Very few women were worried about their "thigh gap" until thinspiration sites made it a thing. Today, thanks to its own hashtag and media proliferation, online commenters regularly disparage women who don't have the elusive thigh gap. Recently, model Robyn Lawley received Facebook comments calling her a "pig" because her thighs touched. In 2016, the A4 waist challenge began circulating on Chinese social media. Women were encouraged to pose with a standard-size piece of paper in front of their midsections to demonstrate that their waists were small enough to be hidden behind the eight-inch width of paper. Such is the power of social media.

Content that feeds the obsession with women's appearance is far from limited to deliberately appearance-focused online communities. Go to any mainstream news website and you won't have to scroll far to find a headline about the latest beauty trick, a female celebrity who lost or gained weight, or something (anything!) about a woman in a bikini. Fox News recently hosted a slide show of female celebrities who showed the slightest deviation from washboard abs and invited viewers to play "Pregnant or Burrito Grande"? In other words, we were encouraged to closely examine each woman's body and decide whether she was in the early stages of pregnancy or whether she simply *ate lunch*.

Many girls and women are now experts at critiquing the images of women used by mainstream advertisers. We know that advertisers are out to persuade, and there's a natural feeling of resistance that comes along with that knowledge. We don't like the idea that some giant corporation is trying to change the way we feel about ourselves or manipulate us into buying things. But when it comes to social media, critical thinking often doesn't even get off the ground. We forget that much of what we see on social media has just as much deliberate persuasive intent as traditional advertising. Kim Kardashian doesn't post sexy selfies just to entertain you. She's building her brand so that you'll buy more of it.

Even social media content without straightforward persuasive intent can still persuade. Maria captured this subtlety perfectly when she described people in her social media feed who seem to post what amount to ads showing their glamorous lives. Every photo of a woman in a bikini that you see online—even if those images are of your friends—sells a message about what it means to be a woman and what's important to our culture. Content posted by friends and relatives might not explicitly be trying to sell you something, but it's still shaping your reality and shifting your values. We're quick to criticize fashion magazines for objectified, unrealistic images of women, but when your friends post these types of images of themselves, that reaction tends to be muted.

Creating the Perfect Self(ie)

Social media sites are almost always highly visual and focused on self-presentation. Girls and women in particular are encouraged to rely on their looks to manipulate the impressions others form of them in the online world. By watching patterns of likes and comments, young girls learn quickly which photos are acceptable to their peers. Tween and teen girls report frantically deleting pictures that aren't getting enough likes, determined to post something better another day. Both boys and girls tend to agree that only girls need to worry about getting likes on their photos—boys would be mocked if they worried about such things. This is one more sign of how much self-objectification is linked with femininity. Monitoring what others think about your appearance is a job that's overwhelmingly assigned to girls and women instead of boys and men.

A few years ago, my lab became curious about online platforms where users upload photos of themselves so that strangers can provide appearance feedback. We surveyed over 200 posters on one such platform—the Reddit thread "Am I ugly?" The women we surveyed were much more likely to say that the experience of posting on "Am I ugly?" and reading the feedback they got left them feeling upset and worse about their appearance than they did before posting—regardless of whether they got negative or positive feedback on the image they posted. It's worth noting that both men and women post on this particular Reddit thread. A man posting might get a taste of the type of appearance feedback women get every day of their lives. But it would just be a taste. Women get this feedback all the time, whether they want it or not, and it has predictable effects on our behavior.

Think about how often you've seen a woman posing for an informal picture with friends, then inspecting it and demanding that it be retaken. That same behavior would seem odd from a man. Not surprisingly, women are much more likely than men to untag photos because they're not pleased with how they look in them. Girls and women are also more likely to edit their online photos using apps like Facetune or Perfect365. Girls as young as twelve report regularly using photo editing not just to smooth out skin or hide acne, but even to change the shape of their faces or size of their eyes or noses.[4] Remember that creating the perfect photo to post also takes time—it's not just a psychological investment.

Posting photos of oneself often becomes an exercise in self-objectification, one that starts alarmingly early. An Indiana University study of profile pictures for teens' social media accounts found that almost half of the girls were either posed in revealing clothing (e.g., showing cleavage) or partially dressed (e.g., wearing a bathing suit, or posing in such a way that clothing was not even visible). A whopping 73 percent of teen girls were posed with what appeared to be a deliberately seductive head tilt.[5]

This trickle-down process by which professional beauty standards are slipping into everyday life is intensified by celebrities who post shots that are not only Photoshopped but also benefit from professional makeup artists and hair stylists. Of course, the seemingly informal selfie taken in the bathroom mirror doesn't reveal any of those behind-the-scene efforts. As evidence of how common photo editing of online images has become, lists of top "Photoshop fails" now regularly include almost as many informal social media images as professionally produced advertisements.

You have more control over your appearance online that you do in the real world. You can cull the best photo from hundreds taken. You can choose lighting and poses designed to flatter. You can filter. You can Photoshop. But the more you see a version of yourself that doesn't really exist, the more foreign the woman in the mirror will feel to you, and the less satisfied you'll become with her. Because now you aren't simply comparing yourself to models, or peers. You're engaging in a particularly devastating form of comparison: your real self versus your created self.

Maria is clear about the fact that social media plays an extensive role in determining how she feels about how she looks. She explained, "If so many photos of me weren't out there, and there weren't so many occasions when a photo was gonna be taken, I would feel completely different about all of this than I do. One of my biggest motivations for looking good and polished, and figuring out my look and all of that, has been so that I look okay in photos. That's been a huge thing."

In high school, Maria had only Facebook to contend with. She spent a lot of time clicking through other people's pictures and accounts, trying to figure out who they were, where they stood in the social hierarchy, and why. She evaluated their appearance and catalogued the ways in which she might win or lose compared to them. But Facebook wasn't as selfie-centric back then. You had to hunt through albums to find many pictures of your peers. By the time Maria was in college, Instagram was on the scene, turning up the volume on beauty sickness.

"I'm very thankful that Instagram was not as much of a thing when I was younger." She laughs halfheartedly. "I think it would have driven me crazy. If I think about some of the girls that were thought of as, like, really beautiful in high school, if I think about if I'd had to look at their Instagrams, I think that would have been, ugh, pretty bad for me. I can definitely see that."

Maria never seemed to like how she looked in pictures. She could quickly identify a fatal flaw that made nearly any photo of her unacceptable—even when her friends swore they couldn't see what she was talking about. She would often think, "Oh no, I'm tagged in some pictures. Ugh, I better go see what they are." The notion that photos of her were out there and that she couldn't control which images of her people did or did not see rankled her.

"I remember being very aware in college of what photos were being taken of me," she explained. "A big thing was people, for formals and stuff, would hire a photographer. And I hated that. Like, absolutely hated that. You know, it was almost like a paparazzi situation, where I was just, like, 'I want to go to this event and have a good time, but I don't want all these photos taken of me.' I would say that has had a big, big effect on my mental health in many ways, and one of the big ways was just always having to feel like I was photo-ready when I didn't want to be."

"You had to always be camera-ready?" I asked.

"Yes," Maria confirmed. "I wanted to look good in the pictures. But my first choice would have been to not be in any of those pictures, if that makes sense."

There didn't seem to be much of a choice, though. Maria explained that "Pictures are gonna be taken absolutely every time you're out." She reflected back to a time when she caught a break from the online-photo frenzy. Maria studied journalism in college and spent a year as the editor of the school newspaper. It was a big job to hold down, especially while taking a full course load. The job left her little time for relaxing with friends or sleeping, but in a surprising twist, it gave Maria what felt like a get-out-of-jail-free card. It released her from that photo pressure.

She's wistful, thinking back on those hectic months. "I feel like the time that I was the most mentally healthy in college was actually when I was the editor of the *Daily*, which is surprising because that's also when I was the most sleep deprived and really not living healthily at all. But I think what was really mentally healthy about it for me was I was always surrounded by the same people, so I didn't care at all what I looked like. At all. I was completely one-track-minded. The only thing I cared about was the *Daily*. And so everything else just sort of went away. And I looked terrible. Like, looking back, I looked awful. I was wearing junk every day, and I wasn't doing my hair, and I looked awful. But I didn't think about it."

"Was there a freedom in not having to think about it?" I wondered.

Maria likes that word for it. "Totally," she agrees. "I just was, like, 'Sorry, guys, like, don't have time for that, I'm too busy for that,' you know? During that period of time, I wasn't going out either, so there were no photos of me for that six-month period. And I was so happy with that. It was like a blessing that I didn't have to worry about that."

I dug a bit more. "Because you wanted to be the journalism superstar, so you didn't have time for silly stuff like that?"

"Exactly," Maria continued. "I think I almost wore that as a badge of honor. I was really big on being, like, 'Oh, I see that you guys are all dressed up, but I couldn't do that because, you know, I was at the *Daily*.' Like, 'I'm too serious to worry about that kind of stuff, good for you guys to worry about it.' That was how I felt."

I'm still stuck on how much those months sounded like an escape for Maria. I tell her, "It sounds sort of freeing, like being busy gave you permission to opt out of those standards. Like you had this ready excuse."

"A hundred percent yes," Maria agreed. "I felt it took me out of the contest. I do feel like I called on that or pulled that out in every situation where it could have been relevant."

It was as though Maria recognized that she, and perhaps every woman, carried the burden of "being pretty" as a job. But she could say, "Sorry, I have this other job. I can't do both, so I don't have to be pretty right now." This is how powerful beauty sickness is: It can leave a woman feeling as if she needs some sort of permission slip to be excused, even temporarily, from the beauty race. I imagine a paper reading, "Sign on the line below to indicate that Maria doesn't have to look pretty today, because she has important things to do."

Body Trolls

When I was young, my grandfather loved to tell his grandchildren the tale of "The Three Billy Goats Gruff." The villain of the story was a hideous troll who threatened to eat the goats, who wanted only to cross the troll's bridge. When my grandfather told the story, he'd employ a growly booming voice for the troll lines, a voice that terrified me at least as much as his telling of the story delighted me. Today, storybook trolls have nothing on the real-life trolls that lurk online, eagerly waiting to rip holes in the self-esteem of women. Though cyberbullying happens to all genders, girls and women are especially likely to be the targets of appearance-related cyberbullying. Whether in the real world or the online world, the words most often used to wound women tend to concern their appearance. It's common knowledge that the ability to bash a woman's looks is a powerful weapon, and often the first weapon in the arsenal of online trolls who target women. Even posts that have nothing to do with appearance are often peppered with commentary about the body shape of the women who write them. And it's not just men who target women in this way. Women do this to other women as well.

In one focus group study, Swedish adolescent girls bemoaned the wide array of appearance-shaming comments they received via social media.[6] Most of the trolling focused on fatness and body size, but also on their breasts, hair, or skin. A group of researchers from Yale and Virginia Commonwealth collected over 4,500 tweets during a single four-hour period that contained the word *fat*.[7] Negative, disparaging, and weight-stigmatizing tweets were significantly more likely to target women than men, and only women were targets of tweets that explicitly linked being heavy with being sexually undesirable (e.g., "I hate imagining fat girls have sex" or "To the abundance of fat girls trying to look sensual by wearing minimal clothing. Don't"). To some extent, every day is fat-shaming day on the Internet, but that didn't stop a group of antifeminists from recently promoting a weeklong online festival of fat shaming with its own hashtag. Again, unsurprisingly, the targets were almost exclusively women. And as usual, the tweets were accompanied by commentary promoting the misguided notion that shaming is the most effective diet plan.

In a recent article titled "Trolls Just Want to Have Fun," researchers from Canada surveyed the personality characteristics of over a thousand Internet users and found that those who engaged in online trolling scored as "everyday sadists."[8] Causing others emotional distress literally makes these people *happy*. Most trolls are willing to hit any target with any insult. The fact that so many choose body-shaming insults when targeting women is telling. It both reflects and encourages the notion that a woman owes it to the world to be visually pleasing to any man whose eyes may fall upon her.

A speech given by Amy Schumer at a Ms. Foundation for Women gala captured the power of body-shaming trolls to pull a woman's thoughts back to the mirror and leave her doubting her own worth. Schumer said, "I want to throw my hands in the air after reading a mean Twitter comment and say, 'All right! You got it. You figured me out. I'm not pretty. I'm not thin. I do not deserve to use my voice. I'll start wearing a burka and start waiting tables at a pancake house. All my self-worth is based on what you can see.' But then I think,

'Fuck that . . . I am a woman with thoughts and questions and shit to say.'"

The Psychological Price of Social Media

In numerous studies of girls and women ranging from junior high through adulthood, high levels of social media use are correlated with:

- greater internalization of the thin beauty ideal.
- more self-objectification.
- more frequent social comparisons.
- higher levels of disordered eating.
- more desire to have plastic surgery.
- greater investment in appearance.
- increased depressive symptoms.

In one study of just over 100 women in the UK, researchers randomly assigned women to do one of three things for ten minutes.[9] They visited either a fashion magazine website, a craft website, or Facebook. Overall, spending time on Facebook put women in a more negative mood. For women who were already prone to making appearance comparisons, scrolling through Facebook for ten minutes also increased their desire to change aspects of their appearance, including their hair, skin, and face. A different study of over 800 U.S. college students found that women who spent more time on Facebook were more likely to think of themselves in objectified ways, and the more they thought of themselves as objects, the more body shame they had and the less sexual assertiveness they reported.[10] Sexual assertiveness is about speaking up with respect to your sexual desires and feelings, including the confidence to say no when you want to say no and avoiding feeling pressured into doing things you don't like. Learning to think of yourself as an object is a high price to pay for scrolling through those newsfeeds.

For Maria, the social-media-fueled beauty pageant became especially potent when she moved to New York to pursue her career. She saw how "put together" her New York friends and colleagues seemed, especially in their beautifully composed Instagram posts, and struggled to feel she could meet that standard. On top of that, thanks to an ongoing stream of social media postings, she had the experience of watching an ex-boyfriend move on with his life in real time. I asked Maria about this period, trying to piece together what happened that so shook her self-image.

Maria took a deep breath, then shared, "I felt like the ground shifted under me. And I was so unhappy. I just felt like I had no structure, I had no infrastructure, I had no support. And I never think, like, 'Oh, I caught a bad break,' or whatever. My conclusion is always, 'Maria, there's something you should be doing better.' Like, 'Step your game up, there's something you could be doing better right now. There are a million things you need to fix.'" Her social media feed was alive with implicit recommendations about all the ways that she could be "fixed."

Maria put a pop culture framework around her experience, telling me, "I felt like it was totally like the beginning of *The Devil Wears Prada*, and I was Andy pre-makeover. I truly was just, like, 'Oh my gosh, I look terrible. I need to figure out what my sense of style is.' I knew my clothes were really poor quality compared to other people's, I really was noticing all of that kind of stuff. So clothes became a really big thing, even though I never thought about that before." Movie buffs might recall that the Andy character was a graduate of

Northwestern's journalism school, making this comparison particularly apt.

The makeover process sounded like a lot of work for Maria, and
I told her so. She enthusiastically agreed, but talked about it as
though it were still happening—not a process she had completed.
"Yeah, it is. And it was totally exhausting. If I told you how much
time I spent, I mean, I already worked out a lot, but I started training for half marathons. I tried to dive into work in different ways.
Tried to figure out hair and makeup and clothes and everything.
Wherever it was that I felt like there was a deficit, I was willing to do
all the work to fix it. I got highlights, and now I go to Sephora and
spend a decent amount of money on stuff. I'm seeing a dermatologist. Just, all of that stuff that I would have thought was stupid or,
like, frivolous before. Like, you know, getting my eyebrows done
really regularly, all of that kind of stuff."

"It sounds expensive!" I added.

"It's really expensive, it's really expensive." Maria laughed. "And
all to create what I see as, like, a New York woman look . . . When I
look at myself in a picture or even in person, what I'm looking at is
little things I would like to tweak. It's like whack-a-mole with different things that I don't want to be happening."

In addition to money, trying to get that "New York woman" look
costs Maria in time. Every morning she's spending around an hour
doing hair and makeup—time that could be used for other things.
She thinks about how her two brothers don't have to give up that
kind of time every morning. She knows it's not fair, but explains,
"Do I like that these are the rules of the game? No. But since they
are, I want to play them to the best of my ability."

Ironically, during the time I was writing this chapter, I was also
arranging to get professional headshots taken for this book jacket.
The experience immersed me in the very environment I usually

work so hard to avoid. I found myself spending a ridiculous amount of time thinking about what I should wear in the headshots. Did I need to dress formally, even though I rarely do so in real life? Should I get new clothes? I worried about makeup. It's standard to have your makeup professionally done for headshots. Would I still look like me with all that makeup on? Would I look better than me? Would I like it if I did?

Despite an incredibly friendly and helpful photographer (who even had her darling dog in the studio to distract and cheer me), I found the whole experience an exercise in beauty sickness. A few days later, after perusing a couple of hundred headshots in order to select my favorites, I'm now aware that my left eye is slightly smaller than my right eye, that I show a lot of my gums when I laugh, and that when I try to smile with my mouth closed, I create a snarl like an angry wolf. I don't want to know about these things (most of which no one would ever notice in real life), and I certainly don't want to waste time thinking about them. I post very few pictures of myself on social media for this very reason. I just don't want to pull any more of my attention or others' attention toward my appearance. It's almost impossible to pose for pictures and examine the result without getting sucked into a detailed rundown of all of your imperfections. I try not to invite that experience into my life any more than I need to.

I didn't realize how much the constant posing for social media pictures had affected young women's behavior until a group of my students taught me about "skinny arm" a few years ago. The term refers to posing with your hand on your hip, elbow jutting sharply out, in order to make your arms look thinner. You've probably seen this pose—it's difficult to take a young woman's picture today without it magically appearing. Once you start looking for the skinny arm pose, you see it everywhere. A friend of mine who is a recruiter sends me pictures of women with skinny arm on their LinkedIn profiles. A relative forwarded me a teenage girl's homecoming pictures. In the group shot taken by a parent in the front yard of one of the girls, every single girl was doing skinny arm. Skinny arm is generally not a natural pose, but it's become the new normal—such is the fear of having one's arms squish out a bit in a picture. Look through group photos of women from ten or twenty years ago. You barely see it. When I was writing this chapter, Northwestern had just completed its annual commencement ceremonies, so my Facebook feed was flooded with photos of smiling graduates. Nearly every young woman I saw in a posed picture was doing skinny arm. There was no such uniformity to the young men's poses. My guess is that few of them even thought about how their pose might affect the perceived shape of their body in photos.

The women who work in my lab demonstrated for me the whole series of mental cues women have learned to go through before their picture is taken. Skinny arm is just the start of things. Turn slightly to the side. Chin out, then down. Slight head tilt. Pop the knee just a bit.

There's nothing wrong with wanting to look nice in pictures. The problem I have with skinny arm is that in addition to glorifying thinness, it represents one more way in which the world of professional models and actresses has infiltrated young women's everyday lives via social media. Those of us who don't pose for a living shouldn't have to be aware of things like skinny arm. Skinny arm is a little reminder that your arms are not okay the way they are. It's one more example of how standards for women's appearance are getting more and more out of hand.

When my lab poses for our yearly picture, I yell, "No skinny arms!" It usually makes people laugh. I like seeing that laughter in the picture a lot more than I would like seeing skinny arms. If we are to have our pictures taken, let those pictures capture something about who we are and what's important to us. Photos need not be one more entry in a battle to be the sexiest or one more attempt to mimic the standards of fashion models.

Immersion in the glossy faux-perfect world of social media can trigger a cascade of psychological pain for women. Maria decided to fight that pain by creating a different kind of online experience for herself and others. She can't opt out of social media entirely, because she needs it for her job. But she does a lot more curating now. She's selective about those she does and does not follow, and tells me she is "unashamed, unabashed about not following people back if I don't really like what they're doing." She's also been able to let go of what she called a "voyeuristic" tendency to see what people were up to and how they looked—even if they were people she didn't care about. She doesn't find the back-and-forth gossip about who got engaged fun or informative anymore. She doesn't follow women on Instagram that she "wants to look like." She has committed to living the life she wants to live, not the one that, as she put it, "creates a cool narrative for other people to consume."

"How do you think of social media now?" I ask Maria.

"I think I realized that there is absolutely no way to know what else is going on in people's lives. And I think people that I at one time maybe put on a pedestal or thought had these really glamorous lives . . . I think I've finally, finally pulled away the curtain now to just be, like, 'No one. That level of glamour is not real. It just isn't. As much as those people might be doing glamorous things, you don't have the full intel on what's going on with them on a daily basis. You don't know what's going on in their minds. You don't know.'" Maria is dead serious when she says this. This was a powerful lesson for her.

Because that lesson was so powerful, Maria didn't just cut down on her social media intake; she also decided to actively do something about the way social media distort our perceptions of what people's lives are actually like. She found many others felt the same way she did about social media, and decided there was a service she could do to "show the difference between what people are presenting in this polished and curated form that is their social feed and what is really going on with them behind the scenes." She got together with a web designer friend and created a site called Cropped (wearecropped.com). Cropped is dedicated to first-person accounts of the lives behind those shiny social media profiles. You can read stories of real people's real lives, with none of the ugly parts cropped out.

"Shattering the illusion of that by putting Cropped together was huge for me," Maria emphasized. "I feel like it has really shown me that what someone's putting on social media isn't their real life, or at least isn't the full picture of their real life—it really took reading enough examples for me to get that. I feel like now that I have read all of those stories, I genuinely think of social media very differently than I did before."

"Do you still find yourself comparing yourself to how other women look?"

Maria takes a breath while she thinks about this. "In the last couple years, I think what I've been working toward is to feel like it doesn't matter. I've been, like, 'Okay, there is no reason why I need to look a certain way.' I try to think through objectively, 'Is there any reason why this matters?'" She pauses for a moment before answering her own question. "I think what I'm concluding more and more is that it really doesn't, and that there are people in my life that care about me and aren't evaluating me based on how I look. Just trying to remind myself that of all my goals, that one's not worthy of that much of my time."

FOUR

The Ways We're Fighting Beauty Sickness Aren't Working

9

Media Literacy Is
Not Enough

A COUPLE OF years ago, a fifteen-year-old named Sarah sent me
an email. Sarah is one of many high school students who have contacted me about their own activism around body image. Hers took
the form of a mini-documentary about how advertising affects
girls. She sent me the YouTube link and asked if I would watch it.
The video begins with a montage of the usual suspects: makeup advertisements featuring inhumanly flawless skin, a fashion ad with
an egregiously Photoshopped model, scenes from the Victoria's
Secret fashion show. Sarah followed these images with a series of
powerful interviews she conducted with girls at her school ranging
in age from twelve to seventeen.

The interviews begin with the girls talking about what beauty
means to them. Their answers would do any parent or teacher proud.
Beauty is on the inside, they explain. It's your personality. It's how
you define it. One girl shares that she finds people most beautiful
when they're sharing something about which they're passionate. If
you stopped at this point in the video, you'd think everything was
fine for these girls. Better than fine, even. But a few more minutes
into the video, it becomes clear that this is not the case.

The next section of the film shows these same girls holding a laptop and looking at a set of advertisements featuring women. The girls catalogue what they're seeing: flawless skin, extreme thinness that somehow coexists with curves. Human Barbie dolls, one explains. You can hear Sarah's voice in the background asking, "How do these pictures make you feel?" The answer, across the board, is "sad." These girls are smart. They know they will never look like the women in these pictures, and quite simply, that makes them sad. They reflect on the times boys at their school called them chubby, fat ass, or too skinny. Times they were called "rocky road" because they had acne, or had their nose made fun of for looking "too Jewish," or were told they'd be hotter if they "just lost a couple pounds." Their openness and vulnerability astound me.

One of Sarah's teachers showed this video to her classes. Several girls cried when they saw it. One boy apologized directly to Sarah for the body-shaming comments he had made. The twenty minutes of that video were a powerful experience for all involved.

The girls interviewed in the film were plenty critical of the images they saw. They called them unrealistic. One made the particularly astute comment that the ideal she sees in ads is "genetically impossible." But do these smart, critical girls still want to look like those airbrushed images? Yes. Without question. This is a fundamental weakness with what's commonly called *media literacy*. In case you haven't heard the term, media literacy in this context involves a few different tasks. The first is cultivating an awareness of the types of messages about women's beauty we're exposed to and what the explicit and implicit arguments in these messages are. Media literacy also entails carefully considering the impact of these messages and supporting activism to build a more healthy media climate. The basic idea is that the more you learn about these media messages, the more you can resist them.

Media literacy is a step in the right direction, but it's never going to be enough to win the battle against beauty sickness. Knowing better is not enough to protect you from the beauty ideal, not when so many other elements of your culture promise you that your happiness is bound to achieving that ideal.

Recently I was asked to give a talk to a group of clinical psychologists. I decided I would trace the path of the research I've conducted on women's body image and give a sort of "lessons learned" presentation. I began that presentation with a section I called, "Back When I Was Young and Foolish."

Back when I was young and foolish (or perhaps just overly hopeful and naive), I truly thought we could soothe women's appearance struggles with a few well-timed doses of media literacy. There was a time when I really thought media literacy would be a panacea. If women just knew how phony these images were, my reasoning went, they wouldn't compare themselves to the models in the images. Without that key social comparison process, the images would lose their power to shape our self-perceptions.

Let me just be clear from the outset: I was wrong.

How Far Can Media Literacy Go?

After watching Sarah's video, I emailed her to ask if she'd like to be interviewed for this book. At the time I talked to her, she was just heading into her junior year at a small school in Hawaii. Sarah is white, but she was born in Hong Kong. She also lived for several years in Canada before recently relocating to Hawaii. Though she hasn't been in Hawaii long, Sarah says it's her favorite place so far.

Like many young women, Sarah has done psychological battle with what she sees in the mirror. When I asked her whether body image was something she had struggled with, she had a story similar to that of many of the women I've talked with. As a kid, she didn't struggle with appearance-related issues. But once she broached adolescence, the mirror caught her eye, and she never quite liked what she saw there. She explained, "I mean, I think every girl, at one point or another, is self-conscious about something. Mostly, I was just completely oblivious to the way I looked up until middle school. And then it was eighth grade when I started feeling a lot more self-conscious, specifically about my weight. And I was never overweight or anything. I just always felt, like, self-conscious about my body."

Sarah's appearance worries changed her behaviors and how she allocated her mental space. When I asked if she was spending a lot of time in front of the mirror at that age, she answered, "Yeah. I was super, super self-conscious. I had really curly hair back then, and I would straighten my hair every single day and I'd wear makeup, and I'd watch what I ate."

"How do you think being so self-conscious affected your life?" I asked.

Sarah's answer was remarkably thoughtful. She summarized the effects for me. "That self-consciousness was kind of the only thing I ever thought about. And it just definitely took a toll on my life. And also I don't like the person that I was in eighth grade, because I was so focused on the way I looked that I also would look at other people and judge those other people. It was such a big part of my life. I just was completely consumed with the way that I looked and the way that other people looked, and I didn't focus as much on who I was or who other people are on the inside."

Ironically, though Sarah now fights the media images that contribute to unhealthy beauty standards for women, when she was very young, she was featured in some of those images. From ages four to nine, while Sarah was living in Hong Kong, she did some modeling. She explained that young white models were in high demand there. Sarah had a vague sense at the time that people thought it was special that she was tall, white, and thin and had big eyes. But she was certainly too young to be cognizant of the issues surrounding the exportation of white beauty standards to other cultures. She's shocked now to think about how she may have unwittingly played a small role in selling Asian women an unattainable white beauty standard.

Sarah looks back on that experience now and wonders how it shaped her. "Now that I'm aware of it, I think, underneath, it definitely did kind of have a bearing on me," she admits. "'Cause it was like, 'Oh, you're so beautiful.' All those different things. And, as opposed to saying, 'You're so smart, you're so kind,' it made me feel like that was the only thing that mattered, and that was where I took all my value from, and that's the only thing that mattered as far as me as a person."

Sarah describes herself as having always been a girlie-girl. She had subscriptions to *Teen Vogue* and *Seventeen*. When she got her first real dose of media literacy in high school, it made a big difference in her life. It was a turning point. She described it as "one of those things where you don't really notice that it's happening in society until you find out about it, and then you notice it everywhere." She further explained: "I would notice around me, that we live in a society where media has such an impact on us. And it completely affects our everyday lives and how we feel about ourselves and other people."

"What kind of an effect do you think media has on you?" I asked her.

Sarah clarified that social comparison used to be a driving force behind her interactions with media. "Well," she explained, "before doing my whole project, definitely I would see pictures of girls in magazines and on TV and it's just—I would look at that and be, like, 'Oh, I don't look like that.'" But things have changed for Sarah since making her video. She says the documentary process shifted the way she perceives media and advertising. She's less passive now when she sees those images. But more important, critiquing those images has pushed Sarah and her friends to wonder why they should even be thinking so much about how they look.

Sarah clarified. "Well, as far as my friends go, I think we would in the past sometimes complain and be catty, and just talk about our appearance and what we didn't like about ourselves. But now we stay away from that, and we don't wanna put an emphasis on that. And so we don't really talk about the way we look."

What do they talk about instead, I wondered. "Everything!" Sarah said, with pride in her voice. I like this idea, that talking less about how they and other girls look leaves them more open to talk about everything else.

I'm not convinced that media literacy has completely turned off the focus on appearance for Sarah and her friends, even if there is some data to back up what she's saying. Later in our conversation, I learned that my instincts were right. Sarah still struggles with beauty sickness at times, despite her best efforts to escape it.

Reviews of media literacy programs delivered to late elementary school and junior high–aged girls show that increasing girls' ability to critically analyze media images of women can reduce weight and shape concerns and internalization of the thin body ideal. Some studies even show that media literacy programs can lead to a long-term decrease in eating-disordered behavior. However, the impact doesn't seem to be so great if programs are delivered only after girls are in high school or college. It might be too late then, with media images having already dug their claws in too deeply. So if you want to spread some media literacy, definitely do it early and often.

But don't stop there, because media literacy programs, even if done early and well, are not enough. Their effects tend to be small.

Media literacy is mainstream today. Spend just a few minutes online and you'll find dozens of sites dedicated to questioning the types of idealized images of women we see every day. Photoshopping exposés are everywhere. Talented filmmakers and activists have spread powerful critiques far and wide. Yet women around the world still struggle mightily with pressures to meet the media-fueled beauty standard. We need to fight back with everything we've got, and that means recognizing that one weapon is not enough.

Caution: Airbrushing Ahead

Activists in the United States, UK, France, Israel, and Australia have been lobbying for a new tool in the battle against unrealistic media images: disclaimer labels that let viewers know an image has been airbrushed or otherwise graphically manipulated. There's a strong history of using these types of labels to improve public health. In attempts to reduce smoking-related disease and deaths, many countries require warning labels on packages of cigarettes. These warnings provide a potent counterpart to cigarette advertisements that feature healthy, happy smokers. The current media ideal for women could also be conceptualized as a public health issue. It contributes to eating disorders, anxiety, and depression in women. Why not warn consumers that what they're seeing isn't real?

The thinking behind this approach is solid. The idea is that if you make the women in those media images seem unrealistic, girls and women won't feel like they're appropriate standards of comparison. If you don't compare yourself to the image, you don't have the opportunity to fall short. But even ideas as appealing as this one need scientific evidence to support them, and unfortunately, the evidence isn't good. Several groups of researchers have examined the impact of adding boxes with text like "Warning: this image has been digitally altered."

Eye tracking studies show that women do notice these disclaimers. The bad news is that the disclaimers don't seem to have a positive impact on women's body image. In fact, some studies show they can actually *increase* some women's desire to look like the model in an image. They can also increase women's drive for thinness, body dissatisfaction, and eating disorder symptoms. One study even found that women who viewed ads with these types of disclaimer labels were *more* likely to self-objectify (compared to women who saw the ads without a label). A particularly worrisome study out of France found that women who viewed these images with warning labels became more prone to negative thoughts.[1] If women saw a disclaimer label while viewing an image of a female model, they were quicker to identify words like *sorrow*, *worthless*, *bad*, and *suicide*. Even more alarmingly, the effect of the label seemed to stick with the image long-term. Two months later, when these women returned to the lab and viewed the same image *without* the disclaimer, the same nasty effect emerged.

A 2015 survey of just over 1,500 British adolescents and adults found widespread agreement among girls and women that airbrushed images have a negative impact on body image (74 percent of girls and 86 percent of women agreed).[2] But this same research also revealed significant skepticism about the ability of warning labels to mitigate the damage these images do. Some women pointed out that without a "before" picture, you can't really know how much or what specific digital alteration has been done, so a simple dis-

claimer label doesn't tell you much. On top of that, research participants argued that the image will always be more powerful than the
words that accompany it—a lesson many advertisers well know and
use to their advantage every day.

How Some Types of Media Literacy May Backfire

A few years ago, my lab was curious about just how media-literate
women are these days. We also wanted to know if being able to critique and argue back against idealized media images of women
actually helps protect against beauty sickness. We created a self-
report measure of how often women engage in critical processing
of the beauty images they see in the media. After collecting data
from hundreds of women and going through several iterations of
our questionnaire, we ended up with three sets of items. The first
set assesses how often women think about how unrealistic/fake
the images they see in the media are. In other words, do they think
about things like airbrushing and Photoshop? The second set is
specifically about the thin ideal we see in the media. It contains
items asking how often women think that the models they see in
media images are too thin or perhaps even suffering from eating
disorders. The final category contains items that directly accuse
advertisers of hurting women with the types of images they use and
questioning why such images are employed. Together, these items
capture the kind of content taught in media literacy programs focusing on images of women.

Several hundred women rated each statement on a scale from 1 (*I
never have thoughts like this*) to 5 (*I always have thoughts like this*). The
table below shows some sample items from each category. You can

complete them yourself and see how you would rate these items on the 1 to 5 scale.

When you see a female model in a magazine, on television, or on a billboard, how often do you have the following types of thoughts?

Images are fake	Models are too thin	Advertisers are hurting women with these images
Nobody's that perfect.	She's too skinny to be healthy.	Images like that make women feel bad about themselves.
You have to have a makeup artist to look like that.	She should eat more.	Images like that make women feel like they have to look perfect.
It takes a lot of camera tricks to make someone look that.	She's way too thin.	Why do models have to be so perfect-looking?
Nobody looks like that without computer tricks.	She looks mal-nourished.	It's not good for women to have to look at things like this.

Based on responses to these items, it's clear that there's a wide range of variability when it comes to how often women critically process the idealized media images of female beauty they encounter. Unfortunately, what isn't clear is that having these types of critical arguments at the ready really helps women. The first two categories of arguments (images are fake and models are too thin) did nothing at all to protect women's body image. High levels of media literacy seemed to provide no immunity whatsoever from beauty sickness. More worrisome was that women who scored high on the last category (advertisers are hurting women) were *less* satisfied with their appearance than those with lower scores. What I suspect is that many of the most powerful arguments women generate in response to these images don't emerge until the psychological damage has already been done. We don't start fighting back until an image has already landed the first punch.

Research from other labs similarly points to the potential for some types of critical processing of beauty images to backfire. Researchers from the Ohio State University and Cleveland State University studied over 200 parents and their adolescent children. They found that when parents engaged their children by questioning the thinness of characters on television or the reality of the appearance of those characters, their children had higher levels of body image disturbance and negative emotions.[3] In a study of high school girls, focusing on the body size and shape of models in magazines was associated with higher levels of eating disorder symptoms, even if those girls were critical of the images.[4] Critically viewing the images requires paying attention to them, and the last thing we should want for our daughters is for that unattainable beauty ideal to get even more attention than it already does.

Another way to look at this issue is to study how women with strong feminist attitudes respond to media images of idealized female beauty. There's no doubt that women who identify as feminist are more likely to have a critical perspective on how women are portrayed in media images. So you might guess that feminism could protect us from beauty sickness. Research reveals a more complicated picture. A Kenyon College analysis of dozens of studies found that feminism does seem to decrease women's tendency to internalize the media-promoted ideal of beauty.[5] That's good news. But there's bad news too. Feminism has little to no effect on how women actually feel and think about their *own* bodies. In other words, feminism makes you less likely to agree with media standards of beauty, but it doesn't seem to help you much when you're standing in front of the mirror.

In chapter 7, I told you about a study I conducted with a couple of hundred college women who wrote down the thoughts they had while looking at magazine advertisements featuring female models. You might remember that long, heartbreaking list of all the negative thoughts those women had as a result of comparing themselves to the models in the ads. What I didn't mention in that chapter is that these same women were *also* very critical of those images. They were media literacy superstars. Over three-quarters listed at least one thought that evidenced critical evaluation of the beauty messages these ads were promoting.

For example, over half of the women complained that the model was unhealthily thin. They wrote things like:

- *The way her hipbone is jutting out bothers me, it looks like she's so thin she's in pain.*
- *It's sickening how models always are so skinny.*
- *This ad would probably do better as an ad for weight loss.*
- *I can see her rib cage, that's gross.*
- *Why are all these damn models toothpicks?*
- *The model's body looks malnourished to me.*

A third of the women in the study noted that the image was "fake" in some way:

- *She must be digitally doctored. That's not natural.*
- *This woman looks really computer edited.*
- *Nobody's eyes are that shade of green.*
- *The more I stare at this ad, it doesn't even seem like that is a real woman but like a painting because her skin is too flawless.*

- *I think she almost looks like a computer image and not a real person.*
- *One word: airbrush.*
- *Nobody looks that good in real life. The picture has been retouched about a thousand times to make her look that good.*

Some even accused advertisers of sending destructive messages about beauty:

- *We always see this type of lady advertised and it makes me and other people feel uncomfortable with our own bodies.*
- *This is a very good ad for bringing down people's self-esteem.*
- *This ad is made to get men off and to make large women feel like crap.*
- *This picture and others like it is why many of my friends starve themselves.*
- *I feel degraded because people normally think we girls need to look like that, but ultimately it's virtually impossible.*
- *If I spent that much time trying to look that perfect, I wouldn't have any time, money, or energy left.*
- *This is not an accurate portrayal of real people. No one is that perfect. So I hope society does not expect me to be that perfect.*

That sounds great, right? Honestly, I loved reading how passionate these women were when it came to deconstructing these images. But if you haven't already guessed, there's some bad news. Generating critical arguments against the beauty standard in the ads did nothing to stop women from comparing themselves to the models, and those comparisons did not have happy outcomes.

Women in the study railed against advertisers for using models who are dangerously thin, then followed those complaints with a fervent wish to be as thin as the model. I believe these women meant it when they criticized the images for promoting eating disorders.

But at the same time, viewing the super-thin models seemed to increase the disgust they felt toward their own bodies, which then made them want to look just like the model they previously criticized for being too thin.

One woman wrote, "This woman looks sickeningly thin. You can see her ribs." She followed those statements with "I wish I looked like this" and "How can I get this look?"

Another wrote, "This body type is unrealistically skinny and her ribs are showing. This picture lowers my self-esteem because I'm not as skinny. I'm not as tan. Should I go to drastic weight-loss programs and tan . . . risking my health? I feel like I want to be like that . . . I wish I was a model. Maybe after seeing this picture I won't want to eat." A different participant wrote, "This ad pisses me off. People should not be that thin! Although, I know I say this wanting to be that thin."

Acknowledging that the images of the models were unrealistic due to retouching also did nothing to dissuade women from wanting to emulate what they saw in the image. One woman wrote, "Nobody's face/skin is that perfect. It's all airbrushed . . . Why can't my skin be that perfect?" Others were even more direct, "I know she has fake eyelashes, but I would like my eyelashes to be like hers." Another wrote, "I'd like my shoulders to be as flawless as her airbrushed ones." Take a moment to let that comment sink in.

When you dedicate a lot of time to critically examining media images, one result is that you spend *more time* processing those images, and you process them more closely. This is likely why those airbrushing disclaimer labels don't seem to be having the influence many hoped they would. The labels cause viewers to focus more on images of women, paying particular attention to perfected areas of the body or face. What sticks in your brain is the image, one more reminder of the beauty ideal. The warning label isn't enough to make that image go away. In fact, it just points out that the ideal is so important and valued that those who created the image put in some extra effort to make it appear.

Knowing better just isn't enough. Being critical of media images of women is a great start, but it's no guarantee that you won't still find yourself stuck in a social comparison downward spiral. Remember that social comparisons are typically automatic and super-fast, often happening at an unconscious level. By the time you start deconstructing an offending image, you've already been hurt by it. Media literacy can help to patch up your psychological wounds, but it can't necessarily prevent them in the first place. It's wonderful for making us more aware of the media climate for women, but that awareness doesn't necessarily reduce the impact of these images. When I talked to Sarah, the high school student and budding filmmaker, she confirmed that although her newfound media literacy has muted the power of the images of women she sees, they still pack a punch.

"I think I've been able to lessen the impact of them, and how they make me feel," Sarah began. "But there's definitely still times, like, even though I've done all this research for this project and watched all these documentaries, there's still times when I'll see something and I know that it's retouched and I know it's fake and that that's an impossible standard that I'm setting up for myself, but it'll still impact me. And I'm not sure how . . . I don't think that ever really goes away."

"So media literacy works a little bit, or it works sometimes?" I followed up. "What do you think?"

Sarah was honest in her reply. "I still am a little bit self-conscious about my weight, but not as much just because I know that the images that I'm trying to be are not real, and that that's unhealthily skinny, and I shouldn't try to look like that."

Sarah explained that even if she has lessened the impact of

pictures on billboards or in magazines, the standard those images have created still lingers. On top of that, a critical approach to media images of women has done little to keep her from comparing herself to "real" women she sees in the flesh—people who aren't retouched.

"So you still end up feeling like you're falling short," I asked, "just from comparing yourself to a friend or someone on the street, or someone like that?"

Sarah sounded a bit resigned in her reply. "A little bit, yeah. But I'm definitely working on that."

Just Walk Away

When I was young, my family would embarrass me by singing lyrics from a song popular in the late 1960s, "Walk Away Renée." Why couldn't I have a cooler song with my name in the lyrics? I wondered. Today I find those song lyrics quite useful when it comes to dealing with many of the media images of women I see in my daily life—much more useful, in fact, than media literacy. When I encounter those Photoshopped images, I walk away. I close the magazine. I look away from a billboard. I switch to a different website. I "flag" an ad. I unsubscribe, unfollow, mute. In whatever manner possible, I walk away. I turn my eyes elsewhere, and I turn my thoughts elsewhere.

Being able to talk back to these images is wonderful. I'm all for informed consumers of media. They make for thoughtful, engaged citizens. But once you know all the media literacy arguments, there's no good reason to let these types of images hit your eyeballs any more than you absolutely need to. Just walk away.

In a recent bout of activism, my lab at Northwestern created a Facebook page to spread this message with respect to a specific context: the checkout aisle of supermarkets and convenience stores. In many areas of our lives, we have some control over the type of media we consume. But in the checkout aisle, you're often stuck seeing the worst types of images on the covers of magazines—images we know are harmful to women. Even worse, the images are often accompanied by all the typical headlines: shape up, slim down, get a bikini body, fix this flaw, hide that problem area. Because you're waiting in line, you're somewhat of a captive audience. Even if it isn't a magazine you would ever buy, you're stuck looking at it. Our solution? Turn it around. Of course, sometimes the back cover is even worse than the front. In that case, you can cover it up with a different magazine that has a healthier image on the cover. The way I think of it, a simple turn-it-around or two can spare the girls and women in line behind you from the negative effects of these images. It's a small, temporary fix. But it can feel pretty empowering. On more than one occasion, I've done some turning it around and gotten a knowing smile or a high five from a woman in line behind me.

The media images of female beauty that surround us are powerful by design. They're worthy of criticism, but they're not always worthy of our time and attention. I'm not saying we shouldn't fight back—we absolutely should. But on an individual level, one of the best ways to sap the power of these images is to limit how often you run into them. Instead of fighting the toxin after it's already entered your system, change what you're consuming. Your attention is extremely valuable. Think carefully about when you want to give it away and when you want to reallocate it to something more worthy.

Since completing her documentary, Sarah has definitely learned to walk away from these images. It's not a perfect solution. You can't always walk away. And walking away today can't necessarily undo the years of media exposure that have shaped and reinforced our ideals. But it's a step in a healthy direction.

Sarah admits that she may never be able to do what one of her favorite teachers recommends: looking in the mirror and believing that she's beautiful just the way she is. But she has another strategy that I highly recommend. Here's how she explains it.

"Just something that I've learned is how you can't put so much emphasis and value on just the way you look, and that you have to dig deeper. And so I've started, or at least tried to, instead of just noticing myself from an outside perspective and what I look like on the outside to other people, to kind of dig deeper and see what kind of person I am, and put more of an emphasis and value on the things that I do and how I act, and my mannerisms and my personality."

"Does it help?" I ask her.

I can almost hear Sarah nodding through the phone. "I think so, definitely. Because it makes you feel better as a person if you focus more on what kind of person you are as opposed to what you look like. And that's all that matters."

"Does it make you happier?"

"Yeah, definitely," Sarah agrees. "I think since I've learned to embrace who I am and focus on who I am on the inside and what I want to be in the future, I'm living a more positive life."

Although she didn't use these specific words, what I hear Sarah saying is that when we walk away from phony, unhealthy, objectified images of women, that leaves us open to walk toward something else. In Sarah's case, media literacy helped her to take those first steps, even if it wasn't enough to get her all the way there. Taking some of the focus off of how she looked let her move her attention to the person she wants to be. She told me she might be interested in pursuing medicine or maybe psychology. She's going to help people. She might just change the world for the better. That's an outcome I can get behind.

10

The Problem with "Real Beauty"

LET ME TELL you a little beauty sickness fairy tale.

Once upon a time, a woman looked in the mirror and didn't like what she saw. She sighed and turned away from the mirror. Then she turned back toward it, as though staring a little harder might change the image looking back at her.

Maybe it was her weight, the shape of her face, or her complexion that caused her disappointment. Maybe she noticed signs of aging and frowned back at them. Maybe her hair wasn't what she hoped it would be. Maybe her clothes didn't hang on her body the way she wished they would. Pick any reason. For this tale, it doesn't matter which one.

Our woman notices a second figure in the mirror behind her. Her good friend has arrived! (Or her partner or her family member. Pick one. For this story, it doesn't matter who.)

"What's wrong?" Good Friend asks, picking up on the frustration in our woman's eyes.

"I'm ugly," the woman replies.

"No!" Good Friend insists. "You're not! You're beautiful the way you are."

The woman looks up at the mirror, a smile brightening her face. "You're absolutely right," she agrees. "I am! I am beautiful."

No matter what she heard after that, no matter what media images she saw, no matter what people posted on Facebook or Instagram, no matter how many diet ads came her way, the woman never again questioned her beauty. She lived happily ever after, secure in the knowledge that she was and always would be physically attractive. The end.

Sound realistic?

Whether you've been the woman in question or the well-meaning good friend, my guess is you can relate to the first part of the story. That part is entirely realistic. The make-believe part comes at the end, when I asked you to accept that simply telling women they are beautiful is the answer to their mirror woes.

If your goal is to help women feel more confident and engage more meaningfully with the world, telling them they are beautiful is not the way to do it. Nonetheless, many well-meaning campaigns rest on the assumption that simply hearing someone tell us we are beautiful can alter our self-perceptions (and our definitions of beauty) so that we're always pleased with our reflections. That's why we see "You are beautiful!" on signs and billboards or written on Post-its stuck to bathroom mirrors. It's why Dove produces those videos that make you cry. It's why the song "You Are So Beautiful" never seems to go out of fashion. There's a part of us that wants to believe those words are a magic balm for women. Unfortunately, in this case, the solution sounds too good to be true because it is. Just as media literacy efforts can sometimes backfire, so can telling women they are beautiful. But there's another way.

* * *

BETH* IS A TWENTY-THREE-YEAR-OLD CHINESE woman who
has been living in the United States for a little over a year. Beth
plans to stay for a couple more years to broaden her work experience
in public relations before returning to China. She was referred to
me by a friend who described Beth as "the single greatest human
being in the entire world, and also hilarious." This friend told me
that Beth had a really different perspective on beauty that I would
find refreshing. She was right.

Beth met me at my office. We chatted around a cluttered table,
occasionally interrupted by students poking their heads in to say
hello. Beth told me she's from what she described as a second-tier
city in China. "It's a small town," she explained, "only six million
people." My eyes widened, and I told Beth we might not consider
that a "small town" in the United States. That's when I was introduced to the wonder that is Beth's laughter, as rapid-fire giggles
bounced around my office. I joined in. You just can't help it when
you're around Beth. Her smile is like sunlight.

"Well, it's only six million," she continued, stifling her giggles.
"Not a lot of people know of it. Definitely second tier."

Beth was young when she first learned the lesson that being
pretty matters to people. When she was around four years old, her
mom, a statistician, would take her to restaurants and let Beth play
with the waitresses while she worked.

"The waitresses would say, 'Oh, you have cute eyes.' Or 'You have
long eyelashes.' Things like that," Beth explained.

"You remember that from when you were so young?" I asked.

"I remember," Beth confirmed. "People would call you cute. They
would say which parts looked similar to your parents. Say, 'Oh, you
got all the good stuff, the good parts!'"

"What are the good parts?" I wondered.

Beth gave me a list. Fair skin, big eyes, long eyelashes, narrow nose, skinny body. I asked her if it was a Western ideal. "Yeah," she said. "Superstar pretty. A face for others to memorize, so we know what beauty is."

Beth was just a little kid when she got those compliments. She took what the waitresses said at face value, telling me, "I bought it. I trusted them." But as Beth got a bit older, she noticed that *everyone* said their friends' little girls were pretty.

When Beth began primary school, she watched as teachers would choose the cutest girls to do things like present flowers onstage to visiting dignitaries. She also noticed that the prettiest girls got "love letters from boys." But Beth wasn't chosen to present flowers onstage. And she didn't get love letters.

You might be bracing yourself for an unhappy ending for Beth, imagining she wound up crying in front of her mirror, wondering why people didn't think she was prettier. But that's not even close to what happened. Beth is not the woman you met in our fairy tale.

The Dove Approach

Here's the message I think most people (or at least most decent people) would want to give to young Beth when she first learned that her features didn't match what her culture considers attractive. They'd assure her that she is beautiful. They might even tell her that she just needs to *see herself* as beautiful in order to feel beautiful, no matter what anyone else says or thinks. In short, they would be tempted to take what I'm going to call the Dove Approach.

We can't talk about the battle against beauty sickness without talking about Dove's "Real Beauty" advertising campaign. I imagine most women have seen at least one of their masterfully produced videos by this point, but you may not have seen the billboards that launched the campaign. These ads were exquisitely shot close-ups of women who, for various reasons, fell outside of the beauty ideal for women. Viewers were asked to consider the woman in the photo and decide which of two boxes should be checked, then vote with their cell phones or at Dove's website. One billboard in Times Square presented a running tally of votes. Is that ninety-six-year-old woman *wrinkled* or *wonderful*? it asked. These billboards were followed by a more widespread campaign featuring women of a variety of body sizes (though, as many noted, all still hourglass-shaped) dressed in plain white underwear and bras, selling cellulite cream that was purportedly tested on "real curves."

Next were Dove's viral videos. The first, "Daughters," was released in 2006. It has the look of a more slickly produced version of the video our talented high school student, Sarah, made. A series of girls of different ages talk about the ways they feel ugly. The second video, the award-winning "Evolution," shows a sped-up version of a woman undergoing makeup, hairstyling, and Photoshopping that turn her from a woman you might encounter on the street to a woman who could exist only in the fantasy world of advertising. "Onslaught," the third video, slams viewers with fast-paced images of models, beauty ads, and scenes of plastic surgery. It exhorts viewers to talk to their daughters about beauty *before* the beauty industry does. More recently, we got to see women describing themselves to sketch artists and to compare the resulting drawings to the (much more attractive) sketches that resulted from strangers describing those same women. The message is sweet. You're more beautiful than you think you are. But just because a message is sweet doesn't mean it's effective.

The mission statement behind the Campaign for Real Beauty sounds laudable: "To make women feel comfortable in the skin they are in, to create a world where beauty is a source of confidence and not anxiety." Dove representatives claimed the campaign was motivated by a shocking statistic from a survey they conducted suggesting that only 2 percent of women around the world believe they are beautiful.

I can believe these ads are well intentioned, even if Unilever, the parent company of Dove, is also responsible for some of the most objectifying, misogynistic ads we've seen. (AXE Body Spray, I'm looking at you.) I can even ignore the unavoidable fact that Unilever would lose massive amounts of revenue if one day women woke up happy with how they looked. I'll even set my cynicism aside and let go of the fact that these campaigns still link beauty with happiness, telling us we could be less sad if we'd simply *accept* that we're beautiful.

Let's take Dove's word for it: Their goal is to make more women feel beautiful. If so, a key question to ask is whether this approach is actually likely to make women feel more beautiful. I don't think it is.

When the "You Are Beautiful" Approach Backfires

One of the sororities on the campus where I teach has been especially dedicated to combating body negativity among young women. A few years ago, they put up a banner in our student center accompanied by a table full of Sharpies. They asked students passing by to write something they loved about their body on the banner. The women running the event were generous enough to let me and my team of research assistants hang around for a few hours and survey women after they wrote on the banner.

For a while I sat in a chair across the room and just watched. I couldn't hear what was being said, but the nonverbals were crystal clear. A sorority member would approach a woman, make her pitch, and offer a marker. I saw dozens of young women stare at the banner and fidget with their Sharpie, the look on their faces like deer in the headlights as they frantically tried to come up with a body part they loved. There were exceptions, of course. I saw one confident young woman walk up to the banner and scrawl, "I love my legs and butt!" Another wrote, "Gotta love my love handles!" But too many young women seemed to struggle to come up with anything at all that they could love about their body. Some gave up, returned the marker, and moved away, clearly worse off for having walked by in the first place. Others approached the banner and wrote a reference to an obscure body part in the tiniest print possible. "I love my cuticles."

Being asked to name what they love about their body was just one more trigger to remind these women of all the things they hate about their body. Thirty percent of the banner-writing women we surveyed said it was difficult to think of a single part of their body they liked. That "love your body" banner was well intentioned, but it looked like it might be having the opposite effect of what was intended.

I designed a simple study to investigate this concern. Young women came to my lab and were left alone at a computer to spend five minutes writing any thoughts and feelings that came to mind. There was just one catch: Every twenty seconds, a chime would sound. We asked half the women to think "I like my body" every time they heard the chime. The other half was asked to think "I am eighteen years old." Immediately after the writing exercise, the women completed a measure of how they felt about their body and their appearance.

Women who were asked to make a simple affirmation—"I like my body"—felt worse about their appearance compared with those who simply acknowledged their age. When we looked at what these women wrote, a pattern became clear. Women who thought to themselves, "I like my body," immediately came up with all the reasons that wasn't quite true. For example, one woman in this condition wrote, "As for thoughts, at this moment I'm wondering what this study is about. I'm guessing it's about body image. Every time I think, 'I like my body,' I'm lying, but I continue to do it anyway when the bell chimes. I try to like it, but I don't really, and every time I have to think that, it reminds me that I don't and I feel all the parts of it that are wrong. I'm pinching my fat right now."

Meanwhile, those who were simply thinking about their age did not focus on how they looked. Instead, they wrote about being bored or hungry or needing to study. They wrote about what they might do later that day. Or about the weather. Very rarely did they even think to mention their appearance.

WHEN WE LEFT OFF WITH Beth, she was just figuring out that she might not be one of the "pretty girls" in her class. Beth explained that this realization didn't disappointment her much, because, as she laughed, "I don't care much!"

Beth explained, "My mom always said beauty is defined by how you behave. How you look is not that important." Beth told me this idea protected her. She used a sports metaphor to elaborate, saying she didn't need to "play offense," trying to convince others she was pretty, because she was "defended" by the idea that looks didn't matter that much.

Though Beth's mom played a key role in shaping Beth's attitudes about beauty, Beth had other positive influences as well. The books Beth read growing up taught her that being kind was the most important thing she could do. How she looked was so far down the list of things that mattered that it didn't capture much of her attention. Instead, she focused on working hard and learning everything she could. She told me, "I studied hard because I'm proud. Proud not because of how I look, proud because of what I achieve."

Beth is no Pollyanna. She's fairly sure that being pretty gives you a leg up, no matter how talented and hardworking you are. She remembers reading a novel when she was sixteen, one that stuck with her for a while. The female protagonist was very smart, and, as Beth put it, "does great work in her field." But the girl was also beautiful. Other characters in the book complimented the girl's appearance. The character's boyfriend was "proud of how she looked," and her friends "showed her off."

For a period of time after reading that book, Beth said she was, "super sensitive about that feeling, that maybe looking good really matters." But the feeling faded.

"Do you ever feel like that now?" I asked.

"Not really." Beth laughs. Once again, I laugh with her.

She explained, "They say, you know, that pretty girls get more opportunities. But for me, I don't believe that much. I feel that why don't you just work harder to be better than those people?" She laughs again, making it clear she holds no ill will toward pretty women who might get unearned favors. In fact, she feels a bit bad for them, noting that they have to spend more money on makeup and spend time keeping up with what's fashionable, two things she has no interest in doing. Instead, Beth told me she tries to focus on "things I'm actually interested in that will make me a better person, to contribute to society."

I'm a little concerned at this point that Beth might be coming off as judgmental, condemning women who worry about how they look. But that's not the case at all. If you were sitting at that table in my office with us while we talked, you would hear that Beth doesn't sound bitter or nasty. She just sounds, for lack of a better word, free. Beth's ability to so wholeheartedly focus on things beyond her looks is unusual among young women. It's undoubtedly been helped along by her generally sunny disposition. But we can all work to be a little more Beth-like.

What I love most about Beth's story is how quickly she pivots from a discussion of how she looks to focus on what she does. This is the pivot we need to be making if we want to fight beauty sickness. Those Dove ads rest on the assumption that all women ought to feel beautiful. Beth is working with a different framework, one that tells her to think less about how she looks instead of trying to think more positively about how she looks. In this respect, Beth is doing exactly what research on this topic would suggest she do.

We Don't Need More Reminders That Our Appearance Is Being Evaluated

Let's get back to those Dove ads by addressing one of their most recent campaigns, "Choose Beautiful." In five different cities, Dove posted large signs above entrances to buildings. Those wishing to enter had to choose between an entrance labeled "beautiful" and another labeled "average." The video begins by showing women who first appear surprised by the options, then sadly walk through the average door. But then one woman confidently enters the beautiful door, and other women follow, heads held high. Some women pull their mothers or daughters through the beautiful door with them.

Take a moment to think about what's happening in the video. Women are going about their days, thinking about whatever they normally think about, when suddenly Dove confronts them with signs that by necessity force them to evaluate how they look. The video also implies that there's something wrong with a woman choosing the average door. Poor thing, she must lack confidence!

Maybe that woman's looks *are* average (after all, it's a statistical impossibility that we can all be above average). There's nothing wrong with looking average. In fact, by definition, most of us *do*. The patronizing notion that women can simply "choose" beautiful also ignores the fact that powerful cultural beauty standards exist. A "Choose Beautiful" campaign is unlikely to be successful in a climate where only certain body types and faces are considered beautiful. These messages don't exist in a vacuum.

Every one of these "you are beautiful" campaigns sends women down the road to thinking *more* about how they look. We don't need any help doing that! Although it was certainly not the intent of the advertisers, these ads actually encourage body monitoring and self-objectification. Body monitoring almost never results in positive outcomes for girls or women, even if it's triggered by a feel-good ad suggesting that all women are beautiful.

The effects of body monitoring can be counterintuitive. For example, most people probably imagine that paying a woman a compliment on her appearance will make her feel better about how she looks. But anything that draws a woman's attention to the appearance of her own body or makes her feel as though her body is being evaluated can result in body shame. Even when commentary on a woman's appearance is meant to be a compliment, it still reminds her that her appearance is being monitored. It brings to mind that ever-present out-of-reach beauty ideal. It's easy to go from "Wow! He thinks I'm hot," to "Wait. He might be looking at my stomach. Is my stomach looking fat in this shirt? How do my legs look? How does my hair look?"

A few years ago, my lab ran an unusual study. It required what psychologists refer to as a confederate. A confederate is typically someone who poses as another research participant but is actually working for the experimenter. For this study, we hired a couple of undergraduate men to work as confederates. They had a simple job. All they had to do was pretend to be another research participant. Then, when the experimenter left the room for a moment, they had to pay the real research participant (always a woman) a compliment. Here's how it went down.

First, the experimenter would give the confederate and the real participant a task to do. The task actually had nothing to do with the hypothesis of the study. It was just a word search designed to keep the participant busy. The experimenter would then leave the room to allow the participant and the confederate to work on their difficult word searches. After a minute, the confederate would "accidentally" drop his pencil. The participant always ended up picking it up for him. In one condition, he would respond with "Thanks, you're nice." In the other condition, he would give the participant a quick look up and down, then smile and say, "Thanks. Hey, that shirt looks good on you." Then they would both go back to their word searches.

The experimenter would return soon after the pencil-dropping exchange and move the participant and the confederate to individual computer stations to "complete the personality measures portion of the study." One of those measures was a test of self-objectification and another was a measure of appearance self-esteem (in other words, how good you feel about how you look). What happened? Not surprisingly, the women who got the appearance compliment reported higher self-objectification. They were more tuned in to how they looked. But more important (and perhaps counterintuitive), they also felt they looked *worse*. In other words, not only did getting checked out and complimented make these women self-objectify, it paradoxically left them feeling less attractive.

A study out of Flinders University in Australia found that even a compliment from a *female* experimenter (in this case, "I was just looking and I like your top, it looks good on you") increased body shame in women.[1] A different study from Syracuse University found that simply telling participants they would be interacting with a male student as part of the study was enough to increase women's body shame.[2]

In a 2016 interview with ballerina Misty Copeland, President Obama made comments that were certainly coming from a kind place. He explained that the Obama daughters have a "tall, gorgeous mom who has some curves," and that it should be helpful to his daughters to see that he appreciates that body type. It's great that he's attracted to his wife and it's lovely to hear that he cares about his daughters' body image. What's not lovely is the subtle implication that his girls should learn to love their own bodies by determining whether men love bodies that look like theirs.

We hear this message all the time. A recent case in point is Meghan Trainor's hit "All About That Bass." Audiences cheered this song as a body-positive anthem. It's hard not to be charmed by the upbeat nature of the song and the lyrics suggesting that "every inch of you is perfect from the bottom to the top." And Trainor did a great job using a pop culture platform to call out Photoshopping nonsense. But here's where it gets a little sticky: In an ideal world, women wouldn't have to worry about being "perfect from the bottom to the top." If we take these lyrics seriously, we also need to consider that the singer is given permission not to worry about her body size only because "Boys like a little more booty to hold," or because she has the "boom boom that all the boys chase." If the narrative of women's body acceptance still hinges entirely on men desiring a particular body type, this isn't really progress. Wouldn't it be better if, instead, we could worry about our intellect and our character and not the appearance of our bodies?

I asked Beth whether her father taught her any lessons about women and beauty. Her answer was unexpected. Beth said she didn't get direct messages from her dad about beauty when she was growing up. But she was certain that he feels the same way about beauty that her mom does. In other words, she was convinced her dad didn't care much about physical beauty.

"How did you know how he felt?" I asked.

"I figured it out by, you know, how he treats my mom. He's not like other men who care about how their women look. She doesn't spend a lot of time looking for clothes or being in style. I think my father's behavior, how he treats my mom, how he respects that, how he never sets those standards where you need to look a certain way, that spreads a message to me that looks aren't important." In other words, Beth's father's actions were more powerful than any "you are beautiful" message could ever be.

I truly appreciate the good intentions behind "you are beautiful" messages. If our option is between ads designed to make women feel vulnerable and unattractive or ads like Dove's that seem designed to make women feel better about themselves, I'd pick Dove's ads any day. But the "you are beautiful" approach, regardless of whether it comes from a good place, can still be problematic. When women suffer as a result of feeling unbeautiful, it's only natural that our tendency is to want to soothe their wounds by telling them how beautiful they are. Yet there are very good reasons to question whether this approach is likely to be effective. Appearance compliments feed a culture that defines women by their looks. Focusing on appearance *at all* turns women toward the mirror—it doesn't matter if that focus is positive or negative.

Women who feel unattractive generally don't believe it when someone tries to assure them that they are beautiful. If beauty sickness were that easy to cure, I wouldn't be writing this book. Instead of messages that reinforce the idea that physical beauty is an essential part of womanhood, we'd be better served by changing the conversation altogether. If we want to improve women's physical and mental health, we need to spend less time talking about beauty and more time talking about issues that matter more. We don't need to talk about beauty in a different way. We need to talk about it less.

Curious about how Beth's desire to be indifferent to beauty played out in her daily life, I asked her if she was a social media user. Beth confirmed that she is active on Facebook and Instagram. "When you're looking at all the pictures, do you notice how the other women look?" I asked.

"A lot of people, a lot of my close friends, they will change things. Use Photoshop to make them look prettier," she noted. "I can show you my social media. You can see I don't do that. I put pictures of myself up. But I show those that really show what kind of person I am."

Beth got excited to show me her photos, pulling out her phone so she could friend me. "I select those pictures when I'm funny or when I'm showing my character," she continued. "Oh, no! My battery died!" she laughed, promising to friend me later. When I received her friend request, I took a look through her photos and understood what she was saying. They feature a lot of grinning and hugging, along with goofy faces. They're not posed. They don't appear to be designed to provoke jealousy or sexual attraction.

Instead, they provide a fun glimpse into Beth's personality and relationships with others.

Beth explained her theory about why women Photoshop their social media images. "This is how people behave when they manage their virtual self. Some will make their virtual self close to their ideal self. Maybe that's why they do that. But for me, I want my virtual self and my actual self to be as close as possible. I don't want to be two different persons."

"What's your ideal self?" I asked.

"My ideal self is being able to step up when the people I love need me. It's fulfilling my responsibility as a daughter or a wife or a mother or an employee. Being good at my job but also doing things for good. That's how I define myself. It's spending time on those things that actually matter." Beth pauses for a moment, then nods, as if putting a final stamp on her statement. "Yeah," she says. "Things that matter."

FIVE

How We
Can Fight
Beauty
Sickness

11

Turning Down the Volume

AT ONE OF the coffee shops I frequent, I often chat with another regular, an older gentleman who likes to talk philosophy. One day as I paid at the register, he was talking with the barista. Inspired by her piercings and tattoos, he made a bold proclamation. "You know," he began, voice booming throughout the cozy shop, "there is no one thing considered beautiful by one culture that is not considered ugly in another. And there is no one thing considered ugly in one culture that is not considered beautiful in another."

"Wrong." I said, perhaps a little louder than I should have. I couldn't restrain myself! I'm a scientist, and he was trespassing in my academic territory without sufficient data to guide him.

"What?" he asked, turning toward me.

"That's not true," I responded, with a matter-of-fact shoulder shrug and a small smile.

"Name one thing!" he demanded. "Name one thing that every culture considers beautiful in a woman."

"Clear, youthful skin," I said.

He frowned at me, unable to argue against that point. The barista stifled a giggle, touched her finger lightly to her youthful, clear-skinned face, and smiled.

I was just getting warmed up. I've been a professor far too long to stop once I've launched into a good lecture. "Shiny, healthy hair," I continued. "Large bright eyes with clear whites. Symmetrical facial features and limbs. An hourglass shape."

"Well, fine," he grumbled. "But you get my point."

And I did get his point. He was basically making yet another "beauty is in the eye of the beholder" argument. We hear these arguments all the time. They're the philosophical gas that drives the engine of Dove's "Real Beauty" campaign. If you accept that physical beauty is defined idiosyncratically by each person in each moment, you can take that idea to the extreme and declare *everyone* beautiful.

As lovely as it sounds, there are two problems with declaring that everyone is beautiful. First, if you do so, you're suggesting that there's really no such thing as beauty. That argument is firmly at odds with piles of scientific evidence to the contrary. Beauty is *to some extent* in the eye of the beholder, but to a meaningful extent, it is not. Second, the "beholder" proposition ignores the fact that part of what drives our perceptions of beauty is rarity. Extreme beauty stands out because it is unusual. If everyone is beautiful, no one is beautiful. I'm not talking about beautiful souls or beautiful dispositions here—I see inner beauty almost everywhere I look. But physical beauty doesn't play by the same set of rules.

A few years ago, a journalist named Esther Honig created an Internet sensation when she paid freelancers from twenty-five different countries to Photoshop a picture of her. Her instructions were simply, "Make me beautiful." In the original picture, Honig's hair is pulled back, she is wearing little if any makeup, and she has a neutral facial expression. When that "before" picture and its accompanying "after" pictures made the social media rounds, the buzz focused on how Photoshoppers from different cultures made such a wide variety of changes to her appearance. One made Honig's lips pink and sparkly, another deep red. Some gave her longer hair. One put a hijab on her.

But the story I saw was not about cultural differences; it was about consistency. When I look at those pictures, I notice that all the Photoshoppers cleared up Honig's skin and evened its tone. They got rid of the dark circles under her eyes and made her features more perfectly symmetrical. They made her eyes look bigger and more defined. In short, they all seemed to be shooting for some version of that list of features I provided to the coffee-shop philosopher. The other alterations they made (makeup colors, hairstyles, jewelry) represented different flavors of icing on the same type of cake.

I want to be clear that I'm not saying culture has no effect on our beauty ideals. It has a massive impact. Culture is why I laugh at the teased bangs and crimped hair I had in the 1980s. Culture is why preferences for lean and muscular versus soft and curvy female bodies have varied across time period and location. Culture can help explain why white women in the United States still go to tanning spas while so many women in areas of Asia and Africa pay substantial sums of money for skin-lightening creams. This book is primarily about culture. But beauty is a complex subject and we do it a disservice if we're not also willing to acknowledge the role of biology in our perceptions of physical attractiveness.

Beauty will always matter to the human mind. Yet it need not matter as much as it does today. Thanks to human evolution, we will likely bump up against our own biology when we try to disentangle ourselves from beauty sickness. But just because we can't turn off our attention to beauty completely doesn't mean we can't turn it down.

Marina*, a white fifty-eight-year-old attorney from Wisconsin with one daughter and one son, knows all about making deliberate choices to turn that volume down. Marina's twenty-five-year-old daughter, Chloe*, sent me a Facebook message in response to a post I made seeking interviewees. Chloe extolled Marina's praises, writing, "My mom miraculously raised me without inflicting any body image issues." I wanted to know more about this feat, so Marina and I arranged a phone interview. She graciously took an hour away from work midday to chat with me from her office.

I generally start my interviews by asking whether the interviewee has any memories from when they were young about how their appearance shaped their lives. For Marina, one specific memory didn't surface. Instead, what stood out were the years of conflict she and her mother had over appearance. The conflict began in adolescence, when Marina gained weight as she developed breasts and hips.

"I was very uncomfortable with that development," Marina explained. "I didn't much understand it, and I knew my clothes weren't fitting properly." On top of gaining some weight, Marina had to contend with the fact that she was unusually tall and long-limbed. She would receive bags of hand-me-down clothing from her cousins, and as she puts it, "I was constantly hanging out of the clothes. My pants would never go down to my ankles and my wrists were always hanging out. And of course my thought wasn't 'These clothes don't fit me.' My thought was 'My body isn't right.'"

"When you were a teenager, what did you think you were supposed to look like?" I asked.

"Barbie?" she answered, with a note of uncertainty. She elaborated. "I was the first generation that grew up with Barbie dolls. I thought I was maybe supposed to look like that, or maybe I was supposed to look like my mother. She was very trim and fit and never a pound overweight. I guess I thought I was supposed to look like that. But I didn't."

Marina's mother had strong ideas about what Marina's body

should look like, and she did not keep those opinions to herself. Marina explained, "She gave me a lot of messages about how I ate too much and I was too big and I was too fat and I should be wearing a girdle or sucking in my belly because I looked too fat. I started getting those messages in adolescence, and that's really the way it's been for the last forty-five years."

Marina's mother is now eighty-six. I asked if she still makes comments about Marina's weight.

"Oh, yeah." Marina draws out the *oh* and laughs sadly. "She does. She's constantly giving me the message that my body isn't right and my body isn't acceptable. I heard that loud and clear for many, many years."

"And I'm assuming from your voice that these weren't subtle comments?"

"Oh, no! Straightforward. 'You're too big. You're not acceptable.'" Marina continued, "Weight is a huge thing with my mother. It's really the lens through which she views most people. If I introduce her to a friend of mine who is skinny, I know she'll think that's just a wonderful person. And yup, if we're out in public, she freely points out women that are overweight and talks about how awful they are. It's a really big thing in her life."

"You said weight is the lens through which your mom views the world. Is it a lens through which you view the world?" I asked.

"No, no, it is not," Marina responds pointedly. "Although, you know, in our body-obsessed culture, where the very small woman is presented constantly through the media as the body ideal that we're supposed to have, I don't know how to escape from that. Even though I have an intellectual understanding of that, I don't know that I am free of it."

I asked Marina whether she thought there would ever be a time when women wouldn't have to worry so much about how they looked to other people. She surprised me with her answer. "You know, when you look at evolutionary psychology, probably not."

"The outlook doesn't look good?"

"It does not look good," Marina confirmed. I laughed a bit, struck that Marina brought evolutionary psychology into the conversation. That's a comment I expect to hear from a college student, not from a lawyer. Marina's pessimism didn't stop her from working to create a body-positive environment for her daughter. It just meant that she was more realistic than most about the forces she was up against.

Evolution and Beauty

The evolutionary account of human beauty that is the source of Marina's pessimism is worth some discussion. But before I get into the science behind an evolutionary framework for beauty, let me tell you about a class demonstration I conduct in my "Psychology of Beauty" course. On the first day of class, I ask students to get into small groups and complete an uncomfortable task. I give each group the same two piles of photos, one of men and one of women, taken from the website hotornot.com. (This is a site where people upload their photos for others to rate.) I ask the groups to sort their two piles from most to least attractive.

The truth is that we silently evaluate the attractiveness of others on a regular basis, but this class activity still makes my students feel awkward. Despite the hours many may spend swiping left or right on Tinder, they're loath to do this sort of ranking out loud. But it's an important exercise because the results are so telling.

There's always a bit of variability in the groups' rankings. For example, among the two from the women's pile who are generally viewed as most attractive, students might disagree about which should get the number one spot. Likewise, some give the guy in the ripped jeans the top spot for men, while others bump him down a notch or two because they think he has an arrogant look on his face. But no one ever puts any of the top three candidates near the bottom of the pile, and no one from the bottom of the rankings in one group ever ends up near the top for another group. With the exception of minor disagreements, students are overwhelmingly on the same page when it comes to physical attractiveness.

Again, I should be clear here that I'm not talking about what people sometimes call *inner beauty*. These rankings were of strangers about whom nothing was known except for their physical appearance. In the real world, we often know things about the character of others that change our perceptions of their attractiveness. You all probably know people who are such jerks that their bad behavior starts to bleed into your perceptions of their attractiveness. Along the same vein, I imagine we all know individuals who started to look more beautiful to us over time as we grew to love and appreciate them. But those minor shifts in perception don't change the fact that our brains are spectacularly good at judging physical attractiveness, and that for the most part, we all seem to be using a similar set of underlying criteria to make those judgments.

For decades now, scientists have systematically examined this type of beauty consensus. Research from the University of Texas examining hundreds of published studies found that we show remarkably high levels of agreement regarding the physical attractiveness of others.[1] For studies where adults in one culture rated the attractiveness of other adults from that same culture, agreement among raters reached the whopping inter-rater reliability coefficient (an index of how much raters agree) of .90. To put that into context, 1.0 would be perfect agreement. Ratings across cultures (where someone from one culture rates the beauty of an individual from a different culture) showed an even more impressive level of agreement; the average reliability coefficient was .94.

You could argue that adults agree about beauty because they've all picked up on the same cultural norms and ideals. But even infants are able to identify attractive faces, and they agree with adults about who is attractive. In one study, nearly 200 six- to ten-month-old infants were shown images that had already been rated for attractiveness by adults.[2] The infants displayed a clear preference for the photos of attractive others, looking at them significantly longer than they looked at photos of less attractive individuals. This pattern held regardless of the race of the person in the photo, and showed up even when the infants being studied were unlikely to have seen many people of that ethnicity. It's really hard to blame advertisers or social media for this type of finding. Babies aren't reading *Cosmo* or scrolling through Instagram.

We evolved to be highly sensitive to human beauty because for millions of years, it served humans well to find beautiful people (in particular, beautiful women) compelling. It's no secret that we still find beauty captivating today. It can literally be difficult to look away from a very attractive individual. When we look at attractive faces, areas of the brain involved in processing rewarding stimuli are activated. Seeing a beautiful face triggers a pleasurable hit of dopamine—not unlike the hit you'd get if you were seriously hungry and someone offered you a piece of cake. But that reaction to beauty isn't just biologically driven. It has also been reinforced by a culture that glorifies the female beauty ideal at every turn.

While there's no doubt that individuals vary in their preferences for different physical traits (some like blondes; others like dark hair, for example), there are a handful of underlying determinants of human physical attractiveness that appear to be universal. Think of it this way: No one asks a makeup artist, "Please, can you make one of my eyes look bigger than the other?" No one pays a plastic surgeon to add wrinkles or blemishes. That doesn't mean we can't look at a woman with wrinkles and find her beautiful. What it *does* mean is that, on average, we will find youthful skin more beautiful than older, less smooth skin. Our culture bears plenty of blame when it comes to our beauty standards for women, but it gets a boost from our biology. If we want to fight the grip beauty sickness has on women, we need to be honest about this boost and where it comes from.

Evolution has had two major effects on perceptions of human physical attractiveness. First, we have evolution to thank for the fact that we are so quick to notice and evaluate human beauty. It takes just milliseconds for us to decide how attractive another person is. It's an incredibly fast process, and it generally proceeds automatically. When we meet someone new, we note how attractive they are as quickly as we note their ethnicity or age or gender. The second major effect is that *what* we find beautiful is partially the product of evolution.

The basic premise of the evolutionary argument for beauty is easy to understand. Any traits that increase humans' ability to successfully reproduce should be passed on to future generations. Throughout human evolutionary history, selecting a healthy mate—one likely to lead to reproductive success—was an important job. Some early humans were better at it than others.

What made some people better at mate selection? When it comes to our male ancestors, one strong possibility is that the most successful were sensitive to and desirous of visual cues that indicated health and fertility in women. We are their descendants, so we've carried on sensitivity to those cues today. You've probably already guessed what those cues are. Early indicators of health and fertility in women are some of the same things we find beautiful today. Youth. Symmetry. Healthy, clear skin. An hourglass shape.

This is where the evolutionary argument loses a lot of people. First, it sounds strange to equate beauty with health, even if we can't ignore the link between age and fertility. But recent studies show that we do rate attractive people as healthier,[3] even if that assessment is happening at an unconscious level. So to some extent, even if beauty isn't a great indicator of health in modern times, we still seem to believe it is.

The argument that sensitivity to women's beauty might have something to do with reproductive success also throws people for a loop. After all, most women today spend most of their fertile years trying *not* to get pregnant. And do you actually know any man today whose main goal in life is siring as many children as possible with as many different women as possible? But here's the thing: That doesn't really matter. When you plot our modern culture on the timeline of human evolution, you can see that the world we recognize today is a tiny blip. It represents a minuscule fragment of time compared to the millions of years of evolution that helped shape us into who we are today.

Evolution in humans is an excruciatingly slow process. As evolutionary psychologists tend to put it, though we live in modern times, we're stuck with Stone Age brains. Our brains have not evolved for the version of life we're living now because the conditions early humans lived in bear little resemblance to our current lives. What we face now is a mismatch between brains that are exquisitely sensitive to variations in human appearance and a culture that drowns us in beauty images.

Overdosing on Beauty

There's no justice in how beauty is handed out or in the quick, automatic responses it generates. But understanding the deep roots of our sensitivity to beauty can help us manage our reactions to it. Evolutionary psychology has a lot to offer in terms of explaining how the current media environment affects us.

We are bombarded with countless images of perfected, unreal female beauty every single day. This ongoing montage of women's faces and bodies leads us to believe that extreme beauty is much more common than it actually is. For millions of years of evolution, no human *ever* saw a face as beautiful as what we might now see *every day* in ads. Now we can hardly escape that type of face. Just as we didn't evolve to be blind to beauty, we also didn't evolve to handle this amount of beauty.

Some evolutionary psychologists use a sugar metaphor when it comes to understanding how the current beauty environment affects us. We evolved to be extremely sensitive to sugar and to desire it fiercely. This makes sense, as for much of human evolution, calories weren't plentiful. It was always worth stocking up before the next famine arrived. Today sugar is cheap and readily available. Just as we find it hard to turn away from beauty images, we find it hard to turn down cookies. In an ideal world, we'd crave vegetables instead of candy, and our health would be all the better for it. But remember, our Stone Age brains weren't made for today. In the same way that abundantly available sugar can have a negative impact on our health, abundant images of highly idealized female beauty can make us sick. This is a key way in which our evolved tendencies interact with our culture to produce beauty sickness.

Our best science suggests that beauty has always mattered. It grabs our attention. That might be natural. But never before in human culture have we been so flooded with nonstop beauty images and messages. There is nothing "natural" about the current beauty climate. Human evolution does not provide an excuse for this culture. But it does underline the importance of changing it. Just as we can choose, for example, to make sugar less available to children in schools, we can choose to make beauty concerns less available to our brains.

I'm not going to try to sell you a bill of goods and tell you that if we just try hard enough, physical beauty will no longer matter in our daily social interactions. The science just isn't there to back up a statement like that. Likewise, I'm not going to take the Dove route, trying to convince you that if you just adopt a different mind-set, you'll always love the way you look.

We cannot completely turn off the part of our brains that makes us so sensitive to human beauty. However, we *can* turn it down. If you don't like the volume metaphor, here's a computer-based comparison instead. Your brain came with a beauty module installed. Maybe you can't uninstall it, but the more you run other programs instead, the less processing power that beauty module will have. So just how do we start to turn down the beauty volume and turn up the volume on things that matter more?

The Lens Through Which You View the World

Marina, the lawyer you met at the beginning of this chapter, is a great example of someone who tries to turn down the volume on beauty. Part of what motivates her is the lingering sting of those comments her mother made about her body. Marina explained, "They were very hurtful. Very hurtful. I had a very low opinion of myself. I really thought I was inadequate and unlovable, for many years." Acknowledging that her mother views women through a lens of physical appearance, Marina works to see the world in a different way.

In psychological terms, I'd use the word *schema* instead of lens. A schema is something that provides organization and structure to your knowledge about yourself and the world. But schemas aren't just passive bookmarks in your brain. They actively guide what you pay attention to and how you fill in the blanks when you have incomplete or ambiguous information. For example, if you have a negative self-schema, you're more likely to interpret an unreturned text message from a friend as evidence that the friend doesn't like you or is mad at you. Someone with a more positive self-schema would be more likely to assume that the friend was just busy or forgetful. In other words, schemas don't just help us organize our experiences; they change the reality that we perceive.

We all have an appearance schema, but they are not all created equal. For some, everything they think they know about themselves or other people is colored by their perceptions of physical attractiveness. Their appearance schema is seemingly always in use, filtering their judgments and insights. For others, their appearance schema gets little use. It lies dormant most of the time.

If your appearance schema is chronically activated, you tend to see the world in terms of looks. When you meet people, you focus on their appearance. If something good or bad happens to you, you assume it had something to do with your appearance. If you watch a movie, you pay more attention to how the actors look than someone with a less busy appearance schema would. The mirror catches your eye more readily. Your thoughts turn to monitoring your own appearance more easily.

Think of a schema like a network, branching into other areas of knowledge. Marina's mother had an extraordinarily strong and well-developed appearance schema. For her, physical appearance connected to personality, to morality, to worthiness, to success. Of all the things she could pay attention to in the world, she tended to fixate on physical appearance. This fixation drove her judgments about other people and influenced her own behavior. It also hurt her daughter over and over again.

Marina made numerous attempts to shift her mother's focus on appearance. She tried to stop her mom's body-shaming comments, especially those targeted at other people. Marina's usual tactic was to tease her mom about the comments, attempting to make her point gently. Other times, she would try taking one of her mom's fat-phobic comments to the extreme in order to point out how absurd it was. If her mother called a woman fat, Marina would sarcastically respond, "Oh, she's so fat? I guess that makes her unlovable and means she has no reason to live. Right?"

"I didn't mean that," her mother would snap back. But another fat-phobic comment would emerge soon after.

When Marina's direct interventions with her mother didn't work, she decided to focus instead on shaping her own life and the life of her children in a healthier way.

When Marina says that she avoids the appearance lens her mother uses, what she's saying is that she doesn't want an appearance-based schema to guide her interactions with the world or her thoughts about herself. She actively fights this schema, breaking the links between her appearance and her sense of self-worth. For example, Marina sometimes finds herself in front of a mirror, bemoaning the size of her hips. When I asked how she pulls herself out of those moments, she explained that she says to herself, "You know what? I'm an acceptable human being, even if I have a couple extra pounds on my hips." She replaces her appearance schema with a different schema that links worthiness to character instead of body shape.

It's not the case that Marina has abandoned all attention to her appearance. That's not the goal of most women, and even if it was, it would be nearly impossible to attain. Marina explained, "I'm not obsessed with my appearance and beauty, but I have an intellectual understanding of how my appearance affects the way other people perceive me. I'm just not totally caught up in that whole world." The process of learning to focus less on her appearance has not been easy for Marina, and she had the benefit of growing up free from social media. If her children endeavor to follow that same path, it will be a much steeper climb for them.

One way Marina has decided to opt out of the beauty race is by refusing to wear makeup. I'm impressed that Marina manages to avoid makeup in a corporate setting, and I tell her so. As a professor, I can eschew cosmetics when I feel like it and get away with few repercussions beyond "Why do you look so tired today?" But Marina's an attorney. Certainly she must face pressure in her workplace to meet traditional feminine beauty ideals.

"Yeah, well," she explained, "I put on some lip gloss. A little bit. That's my concession. But I'm just not going to do a lot of makeup. That's just the way it is."

"Was that a conscious choice at some point?" I probed. "Did you wear makeup and then stop?"

Marina paused to think back. She did wear makeup as an adolescent and into her twenties. She explained, "When I started my job, I thought, 'I should wear makeup, because, you know, I need to look professional.' So I wore a little makeup. And then I got tired of that and just stopped doing it. So yes, I am aware, very aware, that I'm making a conscious choice not to wear makeup, that I'm really violating some of the standards of female appearance. I look around and I just have to say, 'You know what? People seem to think I'm capable of doing this job even though I'm not wearing makeup. So there!'"

Marina acknowledges that some women think it's fun to play with makeup, that it can be about artistic and personal expression. But she also noted that "some women feel that they are really flawed, and if they don't put makeup over the blotchy spot on their face, then they are too embarrassed to leave the house." That's the feeling she doesn't want to feed in herself. And she never wants her daughter to feel afraid to leave the house without makeup. Marina reiterated, "There are various reasons women wear makeup and I don't think any less of them for it. But again, *I'm* not buying into it." I picture her literally putting her foot down. She repeats the sentiment. "I'm just not."

Marina's compassionate and supportive husband helps her along in her attempts to view the world through a lens different from her mother's. His loving attitude and actions speak louder than any words from her mother's mouth ever could. Marina describes her husband's kindness as "healing." If your hips are wide and you still feel loved, that chips away at your mind's association between body size and worthiness. But still, Marina can't entirely escape her appearance schema, because there's always some element of culture ready to reactivate it. She explains, "I go back out into the world and look at the images of women that are presented by the media, and you know, it's hard for me to get away from that old judgmental place and the feelings of inadequacy that are so deep and strong."

The human brain is already too easily turned toward beauty. That's one reason why focusing on appearance more by telling women they are beautiful is unlikely to cure women's mirror woes. After all, your body image isn't just about how satisfied or dissatisfied you are with your appearance. It's also about how much you think appearance *matters*—how invested you are in physical appearance as a guiding construct for your life. If you turn down your appearance schema, you can improve how you feel about yourself without ever changing how you look.

Body image researcher Thomas Cash and his colleagues developed a self-report measure of the extent to which individuals' thoughts about themselves are guided by appearance schemas.[4] Not surprisingly, women significantly outscore men on this measure. A woman who scores high would tend to agree strongly with statements like "What I look like is an important part of who I am." She would tend to disagree strongly with statements like "I have never paid much attention to what I look like" and "I seldom compare my appearance to that of other people." The higher women score on this measure, the worse they tend to feel about how they look, and the lower their self-esteem tends to be. They are also more likely to engage in eating-disordered behavior and compulsive exercise. Seeing the world through the lens of physical appearance isn't good for your psychological or your physical health.

Marina never wanted her daughter's life to be dominated by her appearance schema. "One of my goals as a parent was to raise a daughter who felt fine about herself, who didn't have all the hangups that I did," she told me. "Definitely one of my parenting goals was to raise a healthier daughter than I was. I made every conscious and explicit effort I could not to fall into the same trap that I'd been led into."

"How did you go about doing that?" I asked. After all, this interview came with a ringing endorsement from her daughter. Whatever Marina did seemed to work fairly well.

Marina laughed. "You know, I guess I looked at what my mother did and did the opposite."

When it came to interacting with Chloe about food, Marina

avoided extremes. She didn't insist that her daughter be a member of the "clean plate club," but she also didn't pick on what she ate or tell her she was eating too much. She said, "I tried not to make food a battlefield or an issue, so she would feel free to eat what she wished and really pay attention to her own inner signals about satiety and hunger and learn to eat in a healthier way than I did."

"A lot of people with little girls have the tendency to do a lot of 'Oh, you're so pretty! You're so beautiful!' Did you find yourself doing that with Chloe?" I asked.

"No, no, not so much. In fact, I would really try to consciously say, 'You're such a great friend and you're so nice and you're so smart and you're such a good person.' And so it wasn't specifically about appearance but rather a variety of messages about other qualities."

Marina monitored the types of television programs Chloe watched in order to filter out appearance-focused media or programs that seemed to have the potential to lead to a negative body image. She bought Chloe a subscription to *New Moon,* a magazine for girls that focuses on being an antidote to pop culture's obsession with girls' and women's beauty by spotlighting interesting things girls and women do.

When Chloe came home one day and asked for a Barbie doll, it felt like a critical moment for Marina, whose own body image issues had not been helped by playing with Barbies. Since Chloe seemed to want the doll mainly because playing with Barbies was something her friends did together, Marina and her husband decided not to make a big deal out of it. They had a brief chat with Chloe about the "image" of Barbie and what that meant. Then they purchased Chloe one Barbie and sat back and observed what she did with it. Marina notes with pride, "When her friends were over, she would pull out the Barbie doll and play with Barbie. And then when her friends weren't there, Barbie would go back in the closet and stay there."

Marina knows how hard it is for a girl in this culture to grow up with a healthy body image, so she did her best to provide Chloe with an environment in which the types of hang-ups Marina contended with wouldn't play a role. Marina's approach was not so much focused on saying specific things to Chloe as it was on *not* saying certain things. She didn't comment on food. She didn't comment on weight. She didn't comment on who was or wasn't attractive.

"I tried to be there as a resource without giving her a lot of body messages," Marina explained. "My idea was that she'll get enough of that from the rest of the world. What she needs from her parents is love and support and the message that she's okay. So my husband and I really agreed on that and presented a united 'you're okay' message to her, as well as to our son. We didn't have deliberate conversations about 'You are perfect the way you are,' but we did try to give her the positive message that we loved her. We tried to raise kids who felt good about themselves." Because of the link between gender and body image struggles, Marina seemed to make these choices with Chloe in mind. But she and her husband took the same approach with their son. After all, boys are far from free of body image concerns, and there's no good reason to encourage appearance obsession in any gender. The fact that Marina and her husband were on the same team when it came to this approach also sent a powerful message to their children. It demonstrated that fighting beauty sickness isn't just a job for women. Everyone can and should pitch in.

Part of what drives beauty sickness in women is the feeling that beauty is some essential ingredient to bringing about happiness. How could we not feel that way, when so much of the media we consume reinforces that idea? Marina's mother might have focused on weight so much because she truly thought Marina could be happy only if she was thin.

But it's important to realize that the evidence just isn't there to support the notion that we need to look like models in order to be happy. A study of over 200 college students led by happiness researcher Ed Diener found that the happiest students were no more attractive than those who scored in the average range of happiness.[5] Physical attractiveness does show a *small* correlation with happiness overall, but you have to be careful how you interpret that, because happy people *think* they're more physically attractive than unhappy people. Even acknowledging a small, consistent link between appearance and happiness doesn't mean much for our everyday lives, because there are *much stronger* predictors of happiness than physical beauty. Relationships matter more. A sense of meaning and challenging, interesting work matter more. Likewise, though it's true that your looks affect how others treat you, these effects aren't nearly as large as most people imagine them to be. Beauty-related biases don't make or break you the way we often think they do. Beauty is no magic bullet.

Making the Choice

Body image researcher Eric Stice from the Oregon Research Institute has established a novel approach to improving women's body image and encouraging healthy eating and exercise habits.[6] The intervention relies on a concept called cognitive dissonance. The basic idea behind cognitive dissonance is that we like our thoughts and behaviors to have a good degree of consistency. It feels unsettling, for example, if we think of ourselves as generous and kind but behave in cruel or selfish ways. If that happens, you can resolve the dissonance and regain consistency in several different ways. You could decide that you are in fact a cruel and selfish person. Alternatively, you could modify your behavior to bring it more in line with your self-perception.

Stice's programs harness the power of dissonance to change young women's reactions to idealized media images of women. Young women in the intervention practice openly rejecting the thin body ideal through role-playing and group discussion. For example, in one exercise, the group leader plays the role of someone engaging in severe dietary restraint and the women participating in the program take turns trying to question her beliefs that extreme thinness will lead to health and happiness. In another activity, participants write a letter to a teenage girl who is struggling with body image. In the letter, participants explain the dangers and costs of pursuing the media-fueled ultra-thin body ideal. At the next group meeting, the women read their letters out loud in front of a camera. They can choose to post their video on social media to help spread their message.

Once you've invested significant energy in publicly arguing against this rigid body ideal—as women do in Stice's program— you risk facing cognitive dissonance if you find yourself buying into that same ideal. Over a dozen controlled trials have found that dissonance-based programs result in long-term reductions in young women's internalization of the media-fueled body ideal. To be clear, the evidence does not show that women become impervious to messages about female beauty. But it does suggest that we can attenuate the impact of these messages.

Recent research on these interventions used fMRI (a type of brain imaging) to examine how the brains of women who went through the dissonance-based program reacted to media images of the thin body ideal. The impressive results demonstrated that these women (compared to those in a control condition) showed less activation in the caudate nucleus, an area of the brain associated with reward processing.[7] In other words, it is possible to alter our reactions to burdensome body ideals. We might not be able to eliminate that dopamine hit we get when we look at images of this ideal, but we can dampen it.

However, because we can never entirely erase our sensitivity to human beauty, our best bet isn't just trying to reject unhealthy beauty ideals. What we also need to do is limit our exposure to beauty-focused images and topics. Because our minds evolved to be sensitive to the siren song of beauty and because our culture caters to that vulnerability, it's up to us to turn the volume of that song down.

It's true that most of us need to pay some attention to how we look in order to avoid the social costs that would come from avoiding the mirror altogether. But remember, the trick isn't avoiding beauty completely, it's about putting beauty in its place behind the other things that matter more to you. If you're not sure what those things are, take a moment and make a list of what matters to you. I bet you'll be surprised at how many things top beauty when it comes to your deeply held values.

Putting beauty in its place will mean different things to different women—and that's perfectly fine. My worry is not that beauty plays a role in women's lives. My worry is that it plays an outsized role. As a result, we suffer psychologically and become distracted from things that matter more than how we look.

Putting beauty in its place will involve making a series of conscious choices over and over again until they become habit. Instead of being mindlessly carried along on a wave of beauty sickness, we can practice pushing back against that current. Figure out how hard you're willing to swim to get to a place where your choices around beauty are more consciously made. Imagine swimming until you find a place where beauty pressures are a background trickle instead of an oceanic roar.

WHEN I ASKED MARINA WHAT she sees when she looks in the mirror, she described a tall middle-aged blond woman with gray hairs just starting to make an appearance. Marina reluctantly shared that she hasn't quite accepted her gray hair.

"By not accepting it, do you mean that you're coloring it? Or that psychologically you're having a hard time accepting it?" I followed up.

"Oh, I'm not coloring it. I'm getting used to it. I am who I am, and I am a person whose hair is turning gray, as most people's does at this point in life," she clarified.

I told Marina that one of the young women I interviewed for this book said she imagined that women can finally be free once they let their hair go gray. This young woman looks at women with gray hair and thinks, "Yes, that's the type of older woman I'm going to be. I'll be free."

Marina thought about that idea for a moment. "Boy, I hope she's right," she responded. "I hope I can get to that point someday."

Marina is on her way. Every small step she takes makes a difference. If we all start taking those steps together, those changes could add up to a healthier world for girls and women. The rest of this book is about how to make those choices and take those steps. It's also about how to create a world where it's easier for others to do so as well. This is the kind of world Marina hoped to find for her daughter and herself, but we've got some work to do to get there.

12

Stop the Body Talk

IN 1998, WHEN I was in my first year of graduate school, the television show *Sex and the City* aired an episode called "Models and Mortals." I didn't see it at the time, but in the years that followed, students kept sending me clips of one particular scene in that episode. The four women friends (Carrie, Miranda, Charlotte, and Samantha) sit around Carrie's apartment eating takeout and lamenting the fact that they live in a city with so many models walking the streets. They lob nasty insults in the direction of these nameless models. Miranda claims, "They're stupid and lazy and should be shot on sight," and describes them as "giraffes with big breasts."

Next, these women start a social ritual we know all too well. They share how insecure they feel about their own looks. We learn that Charlotte hates her thighs. Miranda hates her chin. Carrie hates her nose. But then things get interesting. The women look at Samantha, expecting her to join in the body bashing. But Samantha refuses, saying with a shrug, "I happen to love the way I look." Instead of being happy for Samantha and celebrating her confidence, the other women groan and make snarky comments about Samantha having "paid" for her looks.

In case you've lost track of the score, here's what's happened so far.

1. One woman insulted models based solely on how they look.
2. Three of the four women insulted their own appearance.
3. One woman insulted her friend's appearance and two others laughed in support of that insult.
4. Three of the four women were annoyed that the fourth actually likes the way she looks.

It's just a TV show, sure. But the fact that so many women find it relatable points to a sad fact. Instead of using our voices to push back against a culture that undermines our self-worth, too many of us use our voices to attack our own bodies or the bodies of other women.

It's no secret how we got to this place. We grow up hearing a chorus of voices degrading women's bodies. As a result, two things tend to happen. First, we eventually internalize these messages and aim them inward, at ourselves. We call ourselves ugly, we point out our physical flaws, and we tell our friends we look disgusting. Second, we learn that disparaging other women's appearance is a cultural norm, and we join in as though it were a harmless bonding ritual. Both of these types of conversations, those directed at ourselves and those directed at other women, add fuel to the fire of beauty sickness. It's time to consider how we can put a stop to them.

I'd like to introduce two different thirty-nine-year-old women, both of whom have stories to share about the impact of body talk on their lives. Nique has been a regular target of other women's criticisms of her body, while Stephanie* struggles to tamp down the self-directed body talk of the women in her family. Both would benefit from a new set of norms that discourages conversations focused on how women look.

* * *

STEPHANIE IS A THIRTY-NINE-YEAR-OLD WHITE woman living in the suburbs of Phoenix. She's coming up on her ten-year wedding anniversary, and has two young children, one girl and one boy. Stephanie does development work for a local nonprofit and is fortunate to be able to work from home a few days a week. I was lucky to catch her on one of those days when she was enjoying a commute-free evening.

Stephanie has a hard time pinpointing when exactly how she looked to other people became what she calls "a big deal in my brain on a regular basis." But she knows she had abundant reminders throughout her childhood that girls and women need to pay special attention to how they look.

Stephanie has two younger brothers close in age to her. When they were children, the three of them regularly got into knock-down, drag-out fights. One of her brother's go-to moves was scratching Stephanie's face during brawls. Stephanie remembers hearing her grandmother say to her mom, "Don't let him scratch her face! Her skin is too important! She can't have marks. She can't have scars on her face." Young Stephanie didn't fail to notice the double standard. "I could scratch the hell out of his face, you know, no big deal." She wondered why her face mattered more than her brother's.

I asked Stephanie to imagine she could go back in time and ask her grandmother, "Why are you worried about my face getting scratched but not worried about my brothers' faces getting scratched?" How did she think her grandmother would respond?

Stephanie had no trouble imagining her grandmother's explanation. "Oh, I think she would have said the same thing then that she'd probably say now. She'd start right off with, 'A woman's face needs to look like X, Y, and Z,' and, you know, 'A woman's skin should stay smooth and tight.' The implication would be that boys didn't have to concern themselves with such things."

By adolescence, Stephanie had what she described as "an athletic build." She was also on the short side and bustier than the average girl her age. She felt that all the other girls in her school had "longer, leaner bodies" and was angry that she didn't have that same body type. Looking back, Stephanie is certain it wasn't true that all her female classmates were long and lean, but it felt like that at the time.

I wondered where Stephanie got the idea that her build was unattractive. "Why do you think you wanted that taller, thinner body type so much? Somewhere along the line, you had to pick up on this idea that it was more aesthetically pleasing, right?" I asked.

Unlike many women, Stephanie doesn't cite media images or Barbie dolls as the origin of her body ideal. She had a different source for that lesson. She learned it from listening to how the women in her family talked about their own bodies.

The women in Stephanie's extended family are close. Her grandmother, mother, and two aunts spend lots of time with one another. Stephanie calls their group "a little matriarchy." She spent years listening to the types of conversations this matriarchy would have and internalizing their words.

"I heard it on a pretty regular basis from the adult women in my family. They were always so generous to me, but the way they talked about their own bodies and other people's bodies—I heard it all the time. We were always with each other, my mom, my grandma, and my aunts. I learned from the things they said that thinner and taller and leaner were better. Somehow I came away with that."

"Was it that they were criticizing their own bodies?" I asked.

"Oh, yeah. That was a common thing all four of those women do. I heard it most regularly from my own mom, but plenty from my grandma and aunts too. It was a self-deprecating thing." Stephanie notes that none of these women were overweight by any realistic standard. They all had the same basic body type—the body type Stephanie has herself. So when they criticized their own bodies, they might as well have been criticizing Stephanie's.

"It was common to hear, 'Oh, no! My thighs touch!' And with my mom, it was her upper arms. I mean, if I had a dollar for the number of times she complained about how her upper arms looked in my childhood, I'd be a millionaire."

Stephanie explained that the appearance of their bodies was the topic of conversation around which most of their interactions seemed to center. It didn't matter where they were; it was all body talk, all the time. "We'd go to a dinner at somebody's house or a cookout or whatever. It was just the regular topic of conversation." Most often, the women would congregate around the island area of the kitchen. They'd talk about how they were "working on" their own bodies. Or what they didn't like about another woman's body.

There is no doubt that these conversations shaped Stephanie's notion of what her body was supposed to look like. By her first year in high school, Stephanie said she "absolutely hated" her body. "I remember doing aerobics in the family room after school every night, which is absurd for how athletic I was. I remember being obsessed with food, trying my damnedest not to eat carbs. My mom was always obsessed with, like, 'Carbs are the devil.' Even though I loved carbs. I tried, honest to God, I even tried to be bulimic. I was, like, 'I'll just try it,' you know? But I hated throwing up." Stephanie also tried diet pills for a while, but luckily didn't stick with that either.

"Did you think that you could turn your body into a shape you would like? Did that feel realistic to you at the time?" I asked.

"Yes, that's a good question. Yes, I did. I don't know when that went away, when I realized I couldn't. But for a time I was mad at myself that I couldn't make my body turn into what I thought was a good body, into a body type I just actually don't have."

"You would have had to grow taller, right?"

"I would have had to do a ton of things!" Stephanie laughs.

Fat Talk. Ugly Talk. Old Talk.

In 2010, my lab started studying the way women talk about their bodies. At the time, *fat talk* was the typical term used to capture conversations in which women would disparage their own bodies. One of the sororities on campus organized a Fat-Talk-Free Week, but there didn't seem to be much data available about how common fat talk really was, or why women do it.

The first thing we did was survey around 150 college women to ask them basic questions about fat talk. What we learned was eye-opening. First, over 90 percent of these women said they fat talked. Around 30 percent said they fat talked frequently. They had no trouble writing scripts for us to share what it sounds like when they fat talk with their friends.

We got them started with this prompt: *Please imagine that you and a close friend are in class, waiting for lecture to begin. Your friend pinches her thigh says, "Ugh, I feel so fat." Please write the dialogue between you and your friend after the initial comment.* After they finished writing their scripts, we asked these women how they thought their fat talk conversation would end.

Here's an example of what one participant wrote.

Friend: Ugh, I feel so fat.
You: OMG. Are you serious? You are NOT fat.

Friend: Yes I am, look at my thighs.

You: Look at MY thighs.

Friend: Oh, come on. You're a stick.

You: So are you.

How would the conversation end? We would both think we were fat and the other one would still disagree.

Here's another example.

Friend: Ugh, I feel so fat.

You: You're totally not even fat at all.

Friend: I had a donut for breakfast.

You: That's okay. Sometimes you need one.

Friend: I know. I'm just the ugliest person ever probably.

You: That's not true at all, you are totally beautiful.

Friend: Yeah, whatever, I don't want to talk about this anymore.

You: Okay, we don't have to, but you really don't look fat.

How would the conversation end? She feels bad because I gave her a compliment.

When I show these scripts to women, they laugh knowingly. Maybe they've been the fat talker. Maybe they've been the friend who's expected to say, "No, you're not fat! You're beautiful," even though they know they won't be believed. Many have experience being on both sides of these types of conversations. If you look at how the women in this study said these conversations would end, you can see they don't lead anywhere good. Not one woman in our study said she and her friend would feel *better* after this conversation, and many said they'd both end up feeling worse.

Although that particular study focused on college women, fat talk is far from being just a young woman's pastime. In a survey of over 3,000 adult women my lab conducted, around 90 percent said they fat talked at least occasionally. Not until after age sixty did that number really drop off—but only to 83 percent.

There's nothing surprising about the finding that so many women are doing this sort of body talk. If we create a culture that constantly turns women's attention to the appearance of their bodies, of course they're going to talk about their bodies. And because our culture embraces a body ideal few women can approximate, it's no shock that the majority of these conversations are negative.

In a different study my lab conducted, we asked college women to imagine they were with a close female friend in several different scenarios. For each scenario, they wrote a script for the conversation they'd be likely to have with their friend in that setting. One of the settings was trying on clothes in a dressing room. A whopping 76 percent of women wrote dressing-room scripts that included some sort of negative commentary about their own appearance. To put it a different way, fewer than a quarter of our participants could imagine trying on clothes without verbalizing their disappointment with how they looked. They found endless sources of dissatisfaction.

"I'm out of proportion."

"I feel fat and really flat chested. I have no boobs. It's an issue."

"This makes my stomach look huge."

"I have back fat. I know I'm not fat or anything. But I have back fat."

"I have chicken legs. Too skinny."

"I have thunder thighs."

"I'm not getting this. It looks awful. I look awful."

Remember, these women weren't actually *in* a dressing room. But they've been there often enough to know how it goes. Trying on clothes can be downright traumatic. First, there's all that gazing in the mirror. Second, you may be left with the feeling that there's something wrong with your body if the clothes don't fit well, or if an outfit looks not quite the same on you as it did on that size 00 mannequin.

Many of the women my lab has studied tell us they fat talk in an attempt to feel better about how they look. In one study, 60 percent said they thought fat talk was helpful, because it shows them that they're not the only woman who feels bad about her body. In other words, misery loves company. But one of the biggest problems with fat talking is that there's no good evidence it actually does make women feel better. In fact, there's plenty of evidence that it makes women feel worse—whether they're the ones doing the talking or just those listening to it. At some level, women seem to know this. As one seventeen-year-old young woman told us in a survey about fat talk, "Even though my friends try to convince me that I'm not fat, it rarely works and rarely makes me feel better."

The more women fat talk or the more women hear others fat talking, the higher their levels of body shame, body monitoring, and eating-disordered behaviors. A nineteen-year-old student shared her thoughts about fat talk with us after completing one of our studies, writing, "Fat talk makes me more aware of problems with my body that I may not have recognized otherwise." Women who fat talk more also report higher levels of appearance-based comparisons with others.

Fat talk does more than just reflect the body dissatisfaction women already feel; it intensifies it. On top of that, one of the most pernicious effects of fat talk is that it hurts other women who hear it, even if they don't want to engage in it themselves or aren't even part of the conversation. Imagine you're just going about your day when you overhear two women sharing their body concerns. The next thing you know, your attention has been drawn to your own body, and your own body dissatisfaction is triggered. In this way, fat talk can be contagious. It's even worse, of course, when the women you hear fat talking are thinner than you. What are you supposed to think then? As one of our research participants put it, "If you're fat, then I'm humongous!"

To test this notion of fat talk's being contagious, my lab conducted an experiment in which we asked women to come in and discuss magazine advertisements with two other women. What our participants didn't know was that the other two women weren't actual study participants. Instead, they were two of my research assistants, Megan and Heather. The first two ads the three young women discussed were innocuous and didn't feature any models. The third ad featured a model sprawled out in a Calvin Klein bikini. We set up the room so that Megan and Heather would always share their thoughts about the ads first, followed by the actual participant. In the control condition, Megan and Heather made vague comments about the visual setup of the bikini ad that had nothing to do with the model. In the fat talk condition, Megan looked at the model, groaned, and said, "Look at her thighs. Makes me feel so fat." Heather followed suit with "Yeah. Me too. Makes me wish my stomach were anywhere near flat like that." Then it was the participant's turn. What do you think you would say if you were in the study? Would you join in?

In the control condition, when neither Megan nor Heather engaged in any fat talk, none of the participants fat talked. Not one. Even though the women who participated in our study were strangers to Heather and Megan and all had relatively low body mass indices, if they heard Heather and Megan fat talking, 35 percent joined in by adding a "me too" or a disparaging comment about their own body. What's more, across the board, the women who heard Megan and Heather fat talking felt worse about their own bodies afterward, even those who didn't join in. They also felt more shame. But the women who heard the fat talk *and* joined in with their own fat talking got hit the hardest.

Since conducting that first fat talk study, my lab has started using the term *negative body talk* instead of fat talk. Though women's concerns about their appearance often focus on weight, they certainly aren't limited to weight. Other researchers have studied "old talk" among women (conversations that focus on appearance anxiety related to aging) and found such conversations are common across women's life-spans.[1] Though old talk predictably increases as women age, 50 percent of eighteen- to twenty-nine-year-olds admitted that they are already engaging in old talk. Until I read that study, I hadn't even realized how often my friends and I old talk. We complain about emerging wrinkles as though they're a disease. We hunt for gray hairs. I've since made a pledge to myself to curb the old talk. I never want the young women I teach to get the message that women's value decreases with age or that natural signs of aging are something to fear. Women hear enough of this from the culture at large—I certainly don't want to add to those messages.

Stephanie, the thirty-nine-year-old mother of two you met in the beginning of the chapter, heard it all from the women in her family: old talk, fat talk, ugly talk. Any and all appearance flaws were fair game. "Do you remember ever wondering why they talked that way?" I asked.

"No," Stephanie responded, "but I think a lot about it now, because I have kids of my own. Especially because of my daughter. But back then I didn't think about it. In fact, I just thought that it was sort of the norm. That was just what women sat around and talked about."

You Think You're Pretty?

In psychology we talk about two different types of cultural norms. The first is a descriptive norm. When I say that many women struggle to feel happy with their bodies, that's a descriptive norm. It's simply stating what's typical. The second type of norm is an injunctive norm. An injunctive norm carries with it the possibility of social rejection or punishment for those who don't comply. When we look carefully at negative body talk research, there's evidence that women's body dissatisfaction is slipping into injunctive norm territory. In other words, some women seem to feel they're *supposed to* feel bad about how they look. Likewise, some are joining in the negative body talk not because they truly want to, but because they feel they have to in order to fit in with their peer group. Think back to the *Sex and the City* scene. No one was happy when Samantha rejected her friends' negative body talk, expressing confidence in her appearance instead.

Consider the now-classic movie *Mean Girls*. In one scene, the three mean girls are standing in front of a mirror bemoaning their physical flaws. Hips, shoulders, calves, nail beds, pores, hairline. Nothing was good enough. It's clear that the lead character in the film (Cady, Lindsay Lohan's character) feels pressure to join this ritual. But she's not used to it, having grown up in a different culture. Without a readily available physical flaw, she awkwardly lists morning breath as her body problem, much to the annoyance of the other characters. Another noteworthy feature of this scene is that the girls engaging in fat talk are clearly very thin. That's actually consistent with research my lab has conducted. We found that how often a woman fat talks has little relationship with her actual body size. Fat talk is often more about *feeling* fat than *being* fat.

Curious about the idea that women might reject other women

who refuse to engage in negative body talk, my lab designed a study using the *Sex and the City* clip as a stimulus. We brought in women who said they'd never seen the program (so that they wouldn't already have impressions of the characters) and showed them the scene, along with a transcript detailing which character said what. Afterward, we asked them who handled the situation best and why. The women overwhelmingly preferred Carrie's approach. She supported her friends and tried to tell them they were beautiful. She complained about unfair beauty standards for women. But she also managed to malign her own appearance. When asked how Samantha—the body confident character—handled the situation, many women described her as arrogant, vain, and conceited.

Mean Girls also captured this type of scenario perfectly. When one of the mean girls tells Lindsay Lohan's character that she's really pretty, Lohan responds with an awkward "Thank you." Regina George, head mean girl, nastily replies, "So you agree? You think you're really pretty?"

Recently, several young women got press coverage for their decision to respond to unsolicited online body compliments from men by agreeing with them. No matter the platform—Twitter, Tinder, Tumblr—they would politely agree with the compliment. Most of the responses to this social experiment were just what you might imagine. The men would immediately rescind their compliment or call the woman conceited for agreeing with it. When one man told an eighteen-year-old young woman she was beautiful, she responded, "Thank you, I know." His response was to call her a bitch and tell her that being vain won't get her anywhere.

The truth is that women in this culture get a ridiculously mixed set of messages about body confidence. Love your body! But not too much. Be confident! But also humble. Feel comfortable in your own skin, but don't admit it to anyone else, because you might make another woman feel bad. We preach body confidence, but we live in a culture that doesn't quite know what to do with a woman who actually likes the way she looks. It's considered arrogant and even unfeminine. Think of the recent hit One Direction song that made the claim that a woman was beautiful precisely *because* she didn't know she was beautiful. We need to question a culture that tells women they must be beautiful to be loved, but that they shouldn't actually *feel* beautiful or we'll find them conceited.

One result of this paradoxical set of standards is that women are notoriously bad at accepting a compliment. It's one more reason that telling a woman she looks beautiful isn't a good idea for an intervention. More often than not, instead of saying, "Thank you," she'll simply disagree with you. Another result of these conflicting standards is that some women feel they *have to* say negative things about their own appearance in order to be accepted by other women.

Target Practice

On top of women using their words to attack their own bodies, we have all seen the open hostility with which some women evaluate and comment on other women's bodies. Just as girls and women are commonly the targets of commentary about their appearance from men, they're also frequently on the receiving end of negative, judgmental comments about their looks from other women. In adolescent girls and adult women, hearing this kind of body commentary is associated with higher levels of eating-disordered behaviors and with psychological distress.

Though we often think of fat shaming as the primary type of negative body commentary women receive from others, the truth is that few women haven't been targeted for their appearance at one time or another, so narrow are appearance ideals for women. A study of almost 5,000 adolescents in Minnesota found that although over-weight girls received the highest amounts of body-related teasing, underweight girls weren't far behind.[2] Nique, a thirty-nine-year-old married black woman living in the San Francisco Bay Area, can attest to this pattern firsthand. When Nique first emailed me, she said she wanted to talk because she had a story to tell "from the other side of the scale." For much of her life, women have criticized Nique for being too thin.

Nique acts in community theater productions and works a desk job during the day. When we started our phone interview, I asked her how she would describe herself if someone needed to pick her out of a crowd. Nique responded pleasantly, "I would say I'm the perky little brown girl!" Then clarified, "Dark brown. Hershey's cocoa brown."

Nique is tiny. She's always been tiny. When she was young, her mom used to take her out three times a day during the summer to get large milk shakes so she could try to gain weight. When Nique went to college, she said she was the only woman hoping for the "freshman fifteen." But the milk shakes didn't work and the fresh-man fifteen still haven't arrived. Nique says she's tried everything. "I tried Ensure. I tried protein powders. I tried all kinds of stuff, you know? I'm just a little person." I can hear a defensive note in Nique's voice. She's used to trying to justify her body size to other people, even though she shouldn't have to.

Just as our focus on women's bodies opens the door to fat sham-ing, it also explains why the phrase "skinny bitches" has become such a part of our vernacular. Because so many women feel the pain of being heavier than they'd like to be, it becomes easy to lash out against women who are thin. It's open season on the few who manage to meet our culture's body ideal, as though they should be punished for the pain of others despite living under the same set of oppressive requirements. In that same *Sex and the City* episode, we see this too. Miranda actually says she wants to force-feed lard to thin models. Her friends laugh at the idea.

Cruelty toward thin women because they are thin ought to be just as unacceptable as cruelty toward fat women because they are fat. Banning thin women from media images is no more a cure for beauty sickness than banning heavy women from images. Both approaches focus on telling women what type of body is acceptable.

Nique has been a target of skinny-bitch-style cruelty since she was a child. She explained, "People feel like they can just straight up to your face say, 'You're so skinny, it's disgusting.' Or, 'You make me sick, go eat something.' You know, they have no idea how that affects you as a human being, how it affects your self-esteem when somebody just walks up to your face and tells you how disgusting you are."

"People say that to you? To your face?" I gasped. "I wish you could see my face right now," I told Nique, "to see how shocked I am."

"Yeah, well, you know. A lot of people are surprised at that," Nique responded sadly. "I mean," she continued, "a lot of the times they're just joking, but they don't realize that it's not funny."

"What does it do to you to hear comments like that?"

"It hurts, you know?" Nique responded earnestly. "That's kind of a hard thing to hear."

"Do you think they're trying to tell you that you're lucky because you're thin?" I wondered.

"Sometimes it's that. And sometimes, you know, it's straight up just people being mean." Nique says this with a hint of disbelief. No matter how many times this happens to her, she still can't understand why someone would hurt her this way.

When Nique's mom was taking her for milk shakes, it wasn't because she was that worried about Nique's size. As Nique explained it, "She was happy with what I was. But she knew how miserable I was, so she was doing all she could do."

"I know sometimes it's hard to tell the difference," I said to Nique, "but do you think you worried about your body size because other people were saying these mean things to you, or did you really want your body to look different?"

Nique pondered the question briefly and then quietly answered. "I think it was mostly because people were saying things. I'm kind of a tomboy, so I was never into that whole body image thing. I was just busy doing my own thing and having a good time. I wanted to change my body because people made fun of me all the time."

Nique explained that when she was young and out somewhere, any time she'd see two or more people laughing, she assumed they were laughing at her body. "You just always think that anybody that looks at you is thinking something horrible or laughing at you. Which is a pretty awful way to feel all the time."

Words Matter

Nique wants to see an end to the unrelenting conversations around women's bodies. The constant picking on her body by other women has taken a toll on Nique throughout her life. While many women long to be thinner, Nique often wished she had more curves, wishing she weren't "a little-boy-shaped person." While she was growing up, other girls would call her anorexic and bulimic and tell her she was hideous. Strangers still tell her to "eat a burger." Nique thinks people don't realize how much it hurts. "'Girl, you need to eat something. Here, eat this, eat that,' and, 'Girl, what's wrong with you? You're so skinny, ew.' It's just like, I'm sorry. I can't do anything about it."

At the end of our interview, when I asked if Nique had anything she wanted to add, she said she wanted me to let readers of this book know that body shaming happens at both ends of the weight spectrum, and that neither is okay. "More than anything, I really wanna let people know that it really, really makes a huge difference and it impacts young girls when you say things about their body. Maybe you think you're being complimentary when you say, 'Oh my gosh, you're so skinny!' But do you think they don't know that? You might not think that it's a big deal, but it can really crush a person when you're saying things like that to someone who already has low self-esteem. Be careful of what you're saying to people!"

Nique's in a better place now overall. She says she's at a point where she just really doesn't "give a crap what people think" about her body.

"How did you get to that place?" I asked.

Nique stopped to think for a moment before she responded. "I think, first of all, you have to stop looking at what other people think are your faults and just find the things that you like about yourself and focus on those things. I kinda like that I'm tiny. I actually kinda also like that I'm chocolate-covered brown. I have learned to love my skin instead of wishing I was light skinned or wishing I was a different ethnicity. And I also realize that other people, all these people that are saying crap about me, they aren't perfect either. I just, you know, have learned to look at things differently."

"How did you learn to look at yourself differently?" I asked.

"One day I actually looked in the mirror and realized that I had my grandmother's bone structure. And I loved my grandmother. So I was, like, 'Well, you know what? I'm built like my grandmother and that's the way God made me, so this is it, and if you don't like it, don't look at it.'"

Nique's mom also taught her an important lesson. "For her," Nique explained, "it's what's inside. It's what's inside that makes you who you are. You are a beautiful person when you are nice to people and compassionate. I just decided to stop focusing on my body so much and really just focus on the stuff that matters, which is brightening someone's day. Just worrying more about other people and less about yourself."

Nique's method of battling our culture's obsession with how women look would resonate with Stephanie, who tries to teach her daughter, Ella*, similar lessons.

"What are your hopes for Ella in terms of her relationship with her body?" I asked Stephanie.

"I hope that she doesn't think about it that much. I highly doubt she'll be so mentally healthy and oblivious to the rest of society that it doesn't come up. I'm realistic enough to know that it's probably going to seep in here and there, but I just want her to have other things that matter, like I did when I got a little older. Other things that took up my time and my interests to the point where it kind of buries your worries about your body a little bit."

I asked Stephanie whether Ella has ever complained about her appearance. According to Stephanie, Ella doesn't seem to fixate much on how she looks. Stephanie's not sure she can take credit for that, but she is happy with the outcome.

She explains, "We don't talk about bodies at all in front of Ella. There's just no need. Because of how I grew up, I don't want Ella to go through that." I like how Stephanie is trying to break this cycle—refusing to pass on what she received from her mother and grandmother. Stephanie surprises me by saying she thinks it's possible her mom felt the same way—that she wouldn't have wanted to pass on her body worries to Stephanie. I figured with the rampant body talking among the matriarchy, Stephanie's mother wasn't doing much at all to protect Stephanie's body image. But Stephanie explained that wasn't the case.

"She spent so much time trying to build me up. The body talk was never directed toward me. She must have known at some level, like, 'I don't want Stephanie to be concerned about her looks the way that I am.' She's always hated her body. I don't think she wanted me to have that same fate."

Stephanie is sympathetic toward her mom. "She would be just heartbroken if she knew. She would just die if she knew how talking about her own body or other people's bodies had this side effect, this indirect effect on me. I don't think that she ever realized that. And she still does the same thing. Sometimes she refers to herself as fat in front of Ella, and I always look at her and say, 'Please don't. Especially not in front of Ella.'"

I asked Stephanie what she would do if when Ella gets a little older, she came home from school one day and said, "I'm fat. I'm ugly."

Stephanie first answers by joking. With a menacing voice, she says, "I'd ask her who she heard those words from, and then I would go to their house and . . ." But then she laughs. "No, just kidding." Stephanie actually has thought about this scenario, even though she thinks she still has time before it happens. She is certain the moment when Ella comes home with those words on her tongue will one day arrive. She says she knows it will.

"You think it's inevitable that Ella will go through something like that? That at some point Ella isn't going to like the way she looks, and she's going to be upset about it?"

"Yeah, I do. I do." Stephanie agrees. "I think that's a common thing that young girls go through, and I think it's even more tough today because of social media. The ability kids have to do that kind of shaming."

"So how would you handle it?" I ask.

"I guess, two things to do." Stephanie's all business at this point. She's making a plan of attack. "The first is to acknowledge it. I don't want to say, 'Oh no, you're not, stop it, you're beautiful.' That was the route my family took a lot. And I would want to sort of understand where the feelings are coming from a little bit, and say, 'Yes, you know, a lot of people go through this stage where they feel that way.' And then try to move past that and more to the positive side of her, building her up once I've acknowledged her feelings."

Stephanie brings in another important intervention she would make. "The other thing I think is also important, I would also say in that moment, 'You know that it's never nice to say that to anybody else.'" She further explains, "I just think that girls can be brutal to each other, and I don't want Ella to do that. So I would help her through that moment, but also make sure she never does that to anybody else."

Stephanie has no doubts that young girls learn negative body talk—self-directed or other-directed—from the older women in their lives, particularly their mothers. "I shouldn't be saying this three days before Mother's Day"—she laughs—"but yeah! Happy Mother's Day! Stop making your daughters nuts with their bodies! Where do these little girls hear it from? It's just like with me. I stood around the table and heard it from my own family."

Changing the Conversation

When we degrade our own bodies, we send the message to others that it's acceptable for them to do so as well. Negative body talk also sends everyone the message that women should always be worried about their appearance, that hating your body is a "normal" thing for women to do. That's the bad news.

The good news is that we can control the words that come out of our mouths. One of the easiest ways to fight back against our beauty-sick culture is to change the conversations we have about our own appearance. Shifting how we talk about our own bodies is one step toward improving how we think and feel about them. There is no evidence that commiserating with other women about how you look will leave you feeling any better or lead to more positive health behaviors. No one wins when we have these types of conversations, so our best bet is to change the topic altogether. Women have plenty of important things to talk about. We don't have to talk about how we look.

It can take some practice, but you really don't need to give voice to every negative thought about your body that comes into your mind. Imagine if you're talking to a friend and a critical comment about that friend pops into your head. You don't *have* to say it. In fact, I imagine you usually don't. Be as kind to yourself as you would be to that friend.

When you hear other women engaging in negative body talk, don't be afraid to challenge them (gently and kindly) to change the conversation. Lots of this type of talk is just habit—but it's a relatively easy habit to break. When you feel yourself about to launch into negative commentary about your body, take the opportunity to nurture a voice inside your head that can say, "No. I'm not doing that to myself today." Tell yourself, "Today I'm focusing on things that matter more than how I look." Or, "Today I will treat my body with kindness, because it is my home." You won't just be helping yourself, you'll be helping any girl or woman who might have heard your negative body talk.

Changing the way we talk about our own bodies isn't enough, because in a world consumed with how women look, body shaming does not end in the mirror. Instead, we too often target other women in an attempt to build ourselves up. Countless magazine features, television programs, and online forums are dedicated to creating scorecards evaluating how other women look, who wore it best, who needs to gain or lose weight, who has cellulite, and what women of different ages and body types should or should not wear. Women will accomplish more when we see ourselves as members of the same team instead of competitors in a beauty battle. Opt out of these media sources and challenge yourself to talk about things that matter more than how women look.

Words have weight and meaning. They often have more power than we can imagine. Let your words be a balm for a beauty-sick culture. Talk about yourself and other women in a way that sheds light on the things you really care about. Let your words point the way toward a culture that sees women not as objects to be looked at, but as human beings who are ready and able to change the world in remarkable ways.

13

Function over Form

THE LAB I run at Northwestern is populated by bright, engaged students who are dedicated to improving the lives of girls and women. We have raucous weekly meetings, full of laughter and big ideas. That being said, for many years, the results of our studies didn't do much to point us in the direction of making things better. Instead, it seemed we'd spent a significant amount of time focused on describing the problem of beauty sickness, taking snapshots from every possible angle. We gathered data showing why current approaches to battling beauty sickness are insufficient or misguided. We got to know the problem, but we didn't have much to offer in the way of solutions.

Over the years, as I gave talks to groups of young women, the questions I got asked most were "What do we do about this?" and "How do we make it better?" These young women wanted to know how to fight back against beauty sickness in their own lives. They wanted to protect the girls and women they saw suffering. I realized I didn't have much to offer beyond a list of things *not* to do.

Our list of don'ts was long. Don't seek out media that features idealized, objectified images of women. If you must encounter these images, give them as little of your attention as possible. Don't compare yourself to media images of women. Don't fat talk, or even be around it if possible. Don't encourage other women's negative body talk. Don't talk about other women's appearance. Don't choose clothing that requires so much body monitoring it distracts you from what's going on around you. Don't get sucked into appearance-fixated social media. Don't pressure your daughters about their weight. Don't. Don't. Don't.

Until recently, when I got the "What should we do?" question, my honest answer was "We're not sure. We're working on it. We've found a lot of things that make beauty sickness worse, including some things you would guess might improve it. We haven't done a great job finding ways to make it better."

Thanks in part to the inspiration of one of my eternally optimistic students, a few years ago, we started down a more positive road. This particular student didn't want to run a study that would generate more of the type of data that makes us want to hold our heads in our hands and cry. She wanted to try for happy data; she wanted a way to make things better. So we sat in my office and talked through our lists of don'ts until we came up with a promising, proactive do to try.

This chapter is about a realistic alternative to seeing your body as an object for others to look at. It's about reconnecting with your body's breathtaking range of abilities. It's about shifting your perspective on your body from a passive thing for others to evaluate to an instrument that brings your impact to bear on the world around you. It's about embracing function over form.

But first let me introduce you to Amy. I didn't meet Amy until after our "function over form" research was under way, but if I had, she would have been an inspirational case study.

I've introduced a lot of women over the course of this book. Some are still being knocked around by beauty sickness. Others seem to have found a relatively smooth path into a healthier place, with just a few minor detours setting them off course here and there. Amy is difficult to place on that continuum. I think she'd be the first to admit that she has not yet won her battle against beauty sickness. But if you conceptualize beauty sickness as a big brick wall, Amy is the woman who walks up to that wall every day and smashes at it with any tool available. She's a fighter.

Amy is a forty-three-year-old married white woman with twinkling eyes and an endearingly wicked grin. She lives in Chicago, but travels widely for her job as a freelance researcher. Amy and I met at a big Greek restaurant, claiming our table just as the majority of the lunch crowd was filtering out and heading back to work. We had a whole section of the restaurant to ourselves for nearly two hours, as servers cleaned up from the lunch rush and set up tables for dinner. In between bites of hummus and pita, Amy told me of the ups and downs she's faced as a result of living in a culture that can't seem to stop telling her that her body isn't what it ought to be.

Not too long ago, Amy was going through her mom's house in Los Angeles and packing up mementos from when she and her two sisters were kids. She came across a project she'd completed when she was eight or nine years old. Amy and her classmates each wrote an essay and drew pictures, and the teacher bound their work into mini-books for them to take home. Amy opened up her book and looked at her essay. The topic she chose? "I wish I weren't fat."

Amy says she doesn't have a memory from her childhood that wasn't in some way touched by the knowledge that she "didn't look like other kids." She explained, "It wasn't just a matter of being different, but a matter of being wrong. Not looking the way you are supposed to look. Of being kind of a disappointment."

I asked Amy, "If we could go back to around second grade when you made that little book, how do you think second-grade you thought things would be different if you weren't fat?"

Amy seems sad, reflecting on this question. "I thought that people would just like me more. That my mother would be happier, that kids would like me better, and I think I thought that everything would just be easier." The story starts two generations back from Amy, with Amy's grandparents, on her mother's side.

Amy's grandfather was "movie-star handsome" and openly in love with Amy's grandmother, who, as Amy put it, was fat for her entire life. Amy remembers seeing her grandparents' physical affection. Stolen kisses, pats on the butt. She cherished witnessing how her grandfather was "into her grandmother, physically and emotionally." Amy explained that her grandmother's weight "never seemed to slow her down any." She was never on a diet and never seemed particularly worried about her own body size. This setup seems to have the makings of a happily-ever-after. Anyone watching this family might learn the lesson that a large woman is perfectly capable of being happy and active and loved, but that isn't what got transmitted from grandmother to mother.

Instead, Amy's grandmother and her movie-star husband sent their daughter a very different message. Amy points out, "It wasn't even coded, not a bit. In literal words, the message was 'If you are fat, no one will love you. And you need to be thin so you can have a good life, because fat people get discriminated against, and it will be hard to find housing, hard to find a job, and no one will want you. No man will want you.'" Amy rattles off this list of outcomes with a breathless intensity, as though she's reading it directly from some sort of official proclamation.

Somehow, despite the fact that Amy's grandmother appeared on the outside to be both fat and happy, weight obsession became a family tradition. Amy explained, "My grandmother was very obsessed with my mother's weight. And my mother in turn was equally obsessed with ours." Amy says her mother "found it very hard to have fat kids."

Amy grew up in L.A., so as she put it, the "message on the streets" wasn't any better than what she was getting at home. Between her mother and her surrounding culture, Amy learned that "there is a certain way to look and only people who are lazy or morally deficient don't look that way. And it's not only that you can't expect to have love if you are overweight, but you actually don't deserve it."

The pressures Amy faced weren't just about body size, of course. Amy points out that particularly in L.A., having a certain look is a signal of social status. "The beauty standard that L.A. promotes, it encompasses a lot of other elitist expectations, right?" Amy clarified. "Like that if you look a certain way, it elevates you above other people. These are beauty ideals that really only middle- to upper-class women can aspire to."

"Meaning with the right money and time and products?" I asked.

"Right," Amy confirmed. "The money, the time, the products, the right clothing, and the surgeons. So you are proclaiming your class and your worth and many other things by looking the way you are supposed to look."

Amy's mom very much wanted to meet that standard, and wanted Amy to meet it as well. But Amy never got there; she "never felt like she measured up." Amy describes her grandparents as having fought their way into the middle class. They lived their lives feeling precariously perched on the edge of that new social status. Just as Amy's grandmother had struggled to feel she fit in among her middle-class neighbors in the Midwest, Amy's mother played out that same struggle in L.A.

"She was tiny, but she never felt tiny enough," Amy explained. "She always felt like she was this chubby little girl from the Midwest coming to L.A., where all of the mothers were incredibly put-together." Because Amy's mom felt like an outsider, she worked hard to give off the appearance of an insider. Amy says her mom communicated the "sense that it is normal to force yourself to be something you're not, naturally. That's just the price you pay."

Amy and her two sisters all have a heavier body type as their natural state. Amy explains, "If you look at our baby pictures, we were chubby babies. We were chubby children, we were fat teenagers. And we did eat as well as other kids and we were reasonably active, you know? We are just zaftig."

Amy sits back in her chair and grins a bit. "I always joke with people that we were built to keep a man warm on the frozen Russian steppes in a famine. I used to joke with guys, 'Yeah, you think that girl is hot, but halfway through the bad crop failure, she'll be dead and I'm still your wife.' We are built to last."

Don't let the fact that Amy knows how to joke about her body size convince you that everything is sunshine and rainbows. The fact that she happens to have been handed a body type her culture doesn't respect makes her angry. It feels like an injustice that she has to experience over and over again.

"In some ways it is super frustrating." Amy lets herself vent a bit. "I backpack, for example. I go backpack with all of these people and they'll come back and be, like, 'Oh my pants are so loose!' because they've lost five pounds on the trail. I never lose weight backpacking. My body is the most efficient engine, which is great because I don't have to carry as much food as other people. It's great if the potato famine comes, because I will be the last one standing. But on the other hand, it makes losing weight incredibly difficult."

"Does it make you angry?" I ask. I'm a little afraid to hear her response. I'm one of those people she's talking about. I've done nothing to earn my metabolism, and I know it. I had just finished telling Amy how I'd had a terrible, busy quarter filled with minor illnesses and pretty much hadn't exercised in a couple of months. Amy exercises all the time.

"Oh, it makes me furious," Amy says, serious but still kind.

"There is something really unfair about it, right?" I agree.

"Oh, totally unfair. So unfair." Amy shakes her head. "That was really one of the themes of my childhood, this sense of how unfair it was that you are being judged for something you can't help. And nothing, no matter how hard you try—it's never good enough. And it is so easy for others."

Amy is one of healthiest people I know. She's super active. She's pescatarian. When we met for lunch, I ate everything on my plate without thinking much about it. Amy saved a big portion of her food to take home and have for dinner with her husband later. Amy's not blind to this imbalance. She points out, "I didn't eat any differently from my friends. In fact, I ate better than most of my friends. I still eat better than most people I know. I exercise more than most people I know. And I am still fat. I've made more peace with it, but it still rankles."

Amy conjures a combination between a frown and a grin. "I mean, I love you, Renee," she starts, "but for you to tell me that you haven't exercised in two months and you look like that—sometimes it makes me want to kill you. I know you eat like a normal person and that you are not doing anything special to have the body that you do, whereas I am killing myself sometimes just to have a body that's, you know, only forty percent over the body mass index that I am supposed to be or whatever."

Amy's mother kept Amy and her sisters on diets for as long as Amy could remember. "And there were nutritionists and there was fat camp. There was the whole nine yards," she recalled.

"Did you keep up the dieting even when your mom wasn't there to push it?" I asked.

Amy says yes but shakes her head at the same time. "Periodically. I would try. And you know, the problem with dieting is that it is crappy and you are hungry all of the time, and when you are hungry all of the time, you are cranky. It's this rigged game. It's this catch-22 where everybody wants you to be skinny, but my friends would complain that I was cranky because I always had a fucking headache. And I just felt like it was so unfair, because here I am doing what I am supposed to do to make myself look the way I am supposed to look, and then you don't get any support. No one is nicer to you because you are cranky and hungry and can't concentrate. Everybody loves that I am the girl with the good personality and that I am super sweet and fun and bubbly, but it is really hard to be sweet and fun and bubbly when you are starving." The more Amy tried to turn her body into something other people wanted it to be, the less well it served her. It fought back against her attempts at reshaping.

Amy told me a painful story of a time she went to visit a new doctor. The doctor took her blood, did some other tests, but asked nothing at all about Amy's lifestyle, what she ate, or how much she exercised. Yet the doctor did manage to ask Amy, "Have you ever thought about losing weight?"

Amy is still flabbergasted when she recalls that moment. "I just looked at her and said, 'I am a woman in America. What do you think?'"

An Alternative to Objectification

Remember that the essence of objectification is thinking about women's bodies as things to be looked at. Self-objectification, in simplest terms, is what happens when you take that perspective on yourself—you start to see yourself as just a thing to be looked at. But self-objectification is actually a little more complicated, because it doesn't just involve thinking of your body in terms of how it looks, but also failing to think of your body in terms of what it does.

The more you feel your body is a decorative, passive thing, the less in touch you are with the subjective sense of what it can do. When you self-objectify, you might focus so much on your body's shape that you don't even notice things like your energy level or your stamina. As one group of researchers put it, you lose respect for the body as a "physical resource."[1]

Based on this idea, my students and I created a quick online intervention designed to remind women that their body is a physical resource and to create awareness of and gratitude for the things it does. We hoped this intervention might help women feel more positive about their bodies. Over a thousand women ranging in age from eighteen to forty completed our study. After agreeing to participate, they were randomly assigned to complete one set of ten fill-in-the-blank sentences. In the objectification condition, the activity gave women a chance to think positive thoughts about their bodies, but those positive thoughts were always framed in terms of how their body looks to other people. For example:

> The most attractive part of my body is ＿＿＿.
> The sexiest part of my body is ＿＿＿.
> I get compliments on how this part of my body looks: ＿＿＿.
> I love when clothes highlight my ＿＿＿.
> The best looking part of me is ＿＿＿.

This condition basically boiled down to a "you are beautiful" exercise, encouraging women to think of the things they find attractive about their bodies. If you hadn't read chapter 10 of this book, you might guess that completing these sentences would boost women's body confidence. But remember, there are two different ways activities like this can backfire. First, if you're not feeling confident about your body, you may struggle to come up with ways to honestly fill in those blanks. That can leave you feeling even worse than you did before. Second, even if you have no problem coming up with ways to complete those sentences, doing so will still get you thinking about and monitoring how your body looks to other people. That type of body monitoring rarely leads to good outcomes. Instead, it leaves most women perched at the edge of a psychological cliff, with an ocean full of body shame and disappointment below.

Women in another condition in this study got a very different set of fill-in-the-blank sentences—a set designed to help them think about what their bodies do instead of how they look. For example:

> I use my arms to _____.
> My body helps me to _____.
> I love that my body can _____.
> My legs allow me to _____.
> My body feels strongest when _____.

The responses to that set of sentences were the happiest data we've ever seen in my lab. Just reading those responses is a balm for beauty sickness. The women in this condition told us they use their arms to type, cook, carry bags, draw, communicate, and hold people. They told us their body helps them to express themselves, make friends, succeed, feel powerful, travel, move. They said they love that their body can dance, change, run, go to work, take them where they want to go, play sports. One woman's sentence was "I love that my body can live the life I've always dreamed."

After completing the sentence activity, women in the functional condition—the one focused on what their bodies can do—were more satisfied with their bodies compared to women in the objectification condition. Focusing on what their bodies could *do* actually led them to feel better about how their bodies looked. Amy, whom you met earlier, perfectly embodies the take-home message of this study. When she takes her shots at that wall of beauty sickness, she does so by owning the power of her body to *do*, forgetting, at least at times, any issues anyone may have with how it looks.

Amy spent most of her adolescence assiduously avoiding exercise. She explained, "There's this idea that everybody wants you to exercise, but if you go to an exercise class, it's embarrassing and people are not supportive. You know, there is twittering and giggling and it's uncomfortable. You feel like you can't do it, not just because you've never done it before, but because you're fat."

Amy was pretty athletic as a child, but when others looked at her, they didn't see a potential athlete. They just saw a fat girl. Amy described one way this dynamic played out. "My mother used to take us to the park to play baseball, and I was pretty good at baseball. I could catch, I could hit the ball, and yet I was always picked last for teams, even though the week before they'd seen me hit the ball. I got on base more than most people, and yet I still got picked last. Week after week, it would be the same thing. Like it was just too hard for them to remember that I was good at it."

Over time, Amy internalized those attitudes. She'd attribute any type of physical struggle she had to her body size, even when those struggles were normal for kids of any size. If she couldn't do something, she immediately assumed it was because she was fat. When Amy was in elementary school, some classmates asked her to fill in at the last minute on a team that was competing in some type of obstacle course in the school gym. As Amy remembers it, at first she said, "Hell no!" But her friends said they would get disqualified if they couldn't fill the empty spot on their roster, so Amy did it.

"You know, I did the obstacle course. I sucked at it, right? But whatever, I did it. And then I immediately went into the bathroom and locked myself in the stall and started crying hysterically, because I had just humiliated myself in front of my entire school."

Amy's friends didn't understand why she was crying. In their eyes, she did fine. No one had been laughing at her. But Amy told me, "I just had that sense that because I am fat or I am ugly to look at, it must have been so horrible for people to have to watch me go through this thing. Like watching a car wreck. That it must have been like that for other people to watch me. Like it was this unfolding tragedy, right? And I was so upset."

Like many people, Amy first learned to ride a bike as a child. But then she went years without riding again. When she finally got back on a bike as an adult, the first thing she did was crash right into a thornbush.

"A thornbush?" I asked, shaking my head. I'm not a natural bike rider myself, and when I first tried riding as an adult, I promptly ran into a tree and tipped over. I can definitely feel Amy's pain on this one.

"Of course it had to be a thornbush!" Amy added self-deprecatingly.

"I was literally picking thorns out of my ears and hair for like a day. And I got really upset, and my boyfriend at the time didn't understand why I was so upset. I wasn't really hurt, and you know people fall off their bike when they are first learning. I couldn't explain that somewhere inside of me, I was sure that the reason it was so hard for me was because I was fat. If I hadn't been fat, I would be more graceful, I wouldn't have fallen off the bicycle, I wouldn't have embarrassed myself in front of him. For him it was a nonevent. But I couldn't help feeling that it was just one more thing I couldn't do or was bad at because I was fat. Even though there are plenty of fatsoes riding around on bicycles perfectly fine, you know." Amy laughs at the memory now.

Eventually, things started turning around for Amy. Part of what helped was that she had a boyfriend who was so in love with hiking that Amy couldn't help but give it a try. The hiking adventures started out a little rough. Once again, Amy assumed that any challenge she encountered was a result of her body size.

"One time I got really sweaty and I lost a contact lens, and again part of me is thinking, 'Well, if I were thin and fit, that wouldn't have happened.' It's weird because at the same time that I am having these thoughts, the rational part of me knows that it's not true, but it's so deeply ingrained that even though you can rationalize it away, you can't stop those feelings."

Over time, Amy learned that hiking involves some tumbles no matter what your body size. "If I fell down, then that just seemed normal. Sometimes when you are climbing over shit you fall down!" But more important, Amy learned that she liked hiking. And she grew to see her body as something capable, something that could bring her joy.

"It was so much fun. I learned that I could do stuff. I could jump over logs, and I could go up mountains, and I could do all these cool things and I loved it. That was the beginning of feeling proud that I can move my body through space. I can climb up the side of a mountain, and I can climb over big old trees that have fallen down, and I can ask my body to perform tasks and it may not be graceful and it may not be fast, but my body will do it. And it will take me places, beautiful places that I would not be able to see if I didn't ask it to do that, right? Just views of things . . . wild fields full of flowers and tiny mountain streams and things that you don't get to see unless your body takes you there."

As Amy put it, learning to focus on what her body could do didn't just change her muscles. It made her feel "all kinds of strong." Her body was no longer an enemy, keeping her from her goals. It was an ally, moving her toward fun, joyful activities. She got bolder.

A few years after college, Amy was working in San Francisco. On her way to and from her office every day, she passed a jujitsu studio. One afternoon she peeked her head in. A guy in the gym told her the first lesson was free and beckoned her to enter. When Amy countered that she wasn't dressed for it, he said, "No problem, we have a gi in the back." Amy was hooked from that first lesson.

Amy explained to me that jujitsu is a bit like wrestling. You can't do it without touching someone else's body and having them touch yours. "And you can't, you can't worry about whether they're going to accidentally brush a bulge," Amy said, "because you're busy and you're super engaged the whole time."

"You were okay right away with people touching you?" I asked.

"Strangely enough," Amy responded, "I was. Which is odd, because even when dating there were always these little moments where you don't want someone to touch you there because it's squishy or bulgy or rolling out of your pants or whatever. But something so immediately just felt really right about jujitsu. They didn't care that I was fat or short or a girl. All that mattered was that you showed up and you tried and you were you working at it."

Even in the midst of jujitsu, Amy still hadn't shed all of her earlier ways of thinking about herself. One day when Amy was leaving jujitsu, one of her classmates said, "Look, Amy, you're cut." Amy thought he meant she was hurt and searched for an injury. She reenacted this moment at our table, twisting her arm back and forth, scanning it.

"You were looking for where you were bleeding?" I asked.

"Right." Amy laughs. "Then I realized he was talking about muscle definition."

Amy fights to let go of our culture's obsession with women's body size by focusing on what her body does for her. She explains, "I have trained myself to focus on feeling better, being healthier, the physical activities that I do, feeling good about having milestones that are measurable. I climbed down, across, and up the Grand Canyon. That's almost a mile pretty much straight up! And it's hard. Not a lot of people can do that. Even skinny people have trouble with that. I did the Wonderland Trail, which is ninety-four miles and two thousand feet of elevation loss and gain every day for ten days."

"You did that?" I marveled.

"With a pack on!" Amy gleefully responded.

AROUND THE TIME I WAS conducting interviews for this book, a woman from Texas emailed me in response to a TEDx talk I'd given. She said she wanted to put me in touch with entrepreneur Jodi Bondi Norgaard, one of the founders of the Brave Girls Alliance. She thought the two of us would have a lot to talk about, and she was right. Jodi is on the front lines of the battle against beauty sickness, dedicated to creating a healthier world for girls. Amy and Jodi have never met, but I have a feeling they'd be on the same page.

Jodi described two events that left her increasingly sensitive to the fact that girls in our culture could benefit from more focus on what their bodies can do and less on how they look. The first event was positive. Jodi got involved with Girls on the Run, an organization that uses running-based activities to build confidence and life skills in elementary and junior high school girls. The culmination of the program is a 5K race that the girls run alongside their adult buddies. Jodi mentored one young girl as part of this program. She remembers her young partner looking up at her at the end of that race and saying, "Now I know I can do anything." Jodi saw that moment as evidence that teaching young girls to challenge themselves with healthy activities could help build "suits of armor" around them—armor that could protect them as they grew up in a world that would send them so many belittling and objectifying messages.

Not long after that race, Jodi was in a toy store with her daughter, who was nine years old at the time. As they walked up and down the aisles, looking at dolls, Jodi's shock at the types of dolls available for young girls intensified. She picked up one doll she described as having "a belly button ring, crop top, big hair, lots of eye makeup, high-heeled shoes, and big boobs." The tag that shared the name of the doll read "Lovely Lola." Jodi thought to herself, "Oh my god. I don't know one parent that wants their child to look like Lovely Lola, to be inspired to be Lovely Lola." Jodi wasn't just mystified; she was angry. She told me, "I was really sick of the images that are out there, that these young girls are bombarded with. They objectify women. They're telling girls they have to be sexy at a young age. It's not healthy."

Jodi's daughter saw her mom holding the doll and asked, "Mom, is that doll for me to play with?" Jodi thought, "Well, that's a good question."

Jodi actually bought the doll—after assuring the cashier that she was disgusted by it and was just bringing it home to show her husband. And Jodi wasn't willing to just let that doll go. She saw a problem and wanted to do something about it. After two long years of drawing and designing, Jodi created the first in a new line of dolls called Go! Go! Sports Girls. Unlike Lola, who seemed designed to be little more than a sex object, Jodi designed these dolls to encourage young girls to be healthy and active. Jodi based the dolls' measurements off those of her own daughter and her daughters' friends, so that they would have the proportions girls' bodies actually have. She named the first doll Grace, after her daughter. Grace is a young adult now, but Go! Go! Sports Girls are still going. Jodi is determined to get these dolls into the hands of as many young girls as possible.

Going to the Gym, for Better or Worse

Exercise is one of the most powerful ways to feel more connected to and appreciative of your body. It's also good for your brain in ways scientists are just beginning to understand. But as Amy described above, exercise can be a loaded topic for women who don't look like fitness models. If you look at ads for most gyms, it seems that only young, thin women with six-packs are welcome. In real life, of course, gyms are filled with women of all shapes and sizes. But that doesn't mean all women feel equally welcome. In 2016, a woman known primarily for being a Playboy Playmate of the Year felt the wrath of the Internet when she took a picture of a naked woman in the locker room of her gym and shared it on Snapchat. In the post, she covers her face in mock horror and makes a crack about not being able to "unsee" that image. What was the problem with the naked woman in the locker room? Apparently she dared to appear nude without having the type of über-thin, cellulite-free figure we're used to seeing in media images. The poster of that obnoxious (and potentially illegal) photo was roundly condemned and she was banned from her gym, but I wonder how much that helps all the women who saw the story in the news and got one more reminder that even in places that are supposedly designed to foster health, you can't escape the focus on what your body looks like.

Too many fitness settings are unintentional breeding grounds for self-objectification. Mirrors everywhere; tight, revealing clothing; endless opportunities to compare your body to the body of everyone else working out alongside you. On top of that, many fitness settings explicitly focus on the need to achieve the extremely thin, largely unattainable body ideal for women. Posters and other advertisements feature nothing but six-packs and jiggle-free limbs. Instructors exhort women to work for a bikini body—both suggesting that only certain types of bodies are worthy of bikinis and that the motivation for exercise should be to alter one's appearance.

This type of objectified context is nothing but trouble for most women. Numerous studies have shown that women who self-objectify the most are the least likely to stick with exercise. They're also less likely to be able to lose themselves in that pleasant flow state that can come from intense physical activity.

Why you exercise matters. Women who exercise mostly to change how they look are more likely to report body dissatisfaction and disordered eating—no matter what their actual body size.[2] A study of middle-aged women led by researchers at the University of Michigan found that women who exercised primarily to change their body shape not only participated in less physical activity overall, but felt worse when they did exercise.[3] If you exercise only to try to morph yourself into looking like a different person, it's too easy to get discouraged. On the other hand, if you exercise to feel better, to get stronger, to manage stress, or to connect with other people, you're more likely to stay the course.

How we talk about exercise—particularly about women's exercising—too often reinforces the notion that the primary reason to move your body is to whittle it into something more appealing for others to look at. Fitness instructors regularly fall into the trap of berating women's bodies in an attempt to motivate. One of my favorite instructors at the gym I go to jokes about fighting our "hate handles" (instead of love handles) and tells the women in her classes, "You want lipo? I'll be your doctor. I'll be your surgeon." I love this particular instructor. She's energetic and funny and teaches great classes. But I don't want to hear the word *hate* anywhere near my body when I'm exercising. And I can't imagine a scenario where it's a good idea for women to be thinking about liposuction in a setting that's supposed to promote health.

My lab recently surveyed over 300 women from around the United States who regularly take group fitness classes. We asked them what types of motivational comments their instructors made, and what their most and least favorite comments were. Then we surveyed over 500 U.S. group fitness instructors, asking them what types of motivational comments they use when they teach classes. What we found was evidence of a clear disconnect between women who take these classes and the instructors who teach them. First, women reported hearing appearance-focused commentary from their instructors (things like "Blast that cellulite!" and "Get that bikini body ready!") much more often than instructors admitted to making those comments. Not only did women report regularly hearing this type of commentary from their fitness instructors, they also told us that they didn't appreciate it. When asked to list their least favorite motivational comment, over half of the women listed something related to appearance pressures or weight loss. For example, one woman wrote that she hated hearing "It's almost bikini season!" When we asked her why, she responded, "It makes me more self-conscious about my body in a bikini. Even if I am comfortable with my body, I'll question it the second I hear that phrase."

Other women wrote of their distaste for how instructors focus on weight loss, as though that was the only possible reason for exercising and the only appropriate metric of fitness success. One woman said that "Get rid of that belly fat!" is the comment she least wants to hear from a fitness instructor because, as she put it, "I'm ALWAYS going to have belly fat. I've been at a place where I run six miles a day and eat really well and I still have a bit of belly fat. But so what?"

We also asked our fitness participants and instructors to complete a measure called the Reasons for Exercise Scale.[4] This measure asks you to rate the extent to which your exercise is motivated by three different types of factors—appearance, health/fitness, or enjoyment/mood improvement. Instead of asking instructors to complete the measure for themselves, we asked them to complete the measure from the perspective of the women who take their classes. Once again, we saw a notable disconnect. Group fitness instructors significantly overestimated the extent to which women in their classes were motivated by appearance and underestimated the degree to which women were motivated by fitness.

If instructors believe the women in their classes are exercising primarily to get closer to the cultural body ideal for women, then of course we're going to hear those beliefs reflected in the words they use to talk to women about exercising. But I would argue that women don't need any reminders that our culture expects them to be thin. And we do a disservice to women when we focus on working out to look good instead of working out to feel good. We're much better off focusing on how exercise makes us feel instead of how it makes us look.

Feeling Your Body Instead of Seeing It

Let's get back to Amy, our mountain-climbing, jujitsu-fighting forty-three-year-old. I loved hearing Amy's stories of the moments in which she felt she was successfully beating back our culture's obsession with women's body size and her mom's reinforcement of that obsession. Big or small, those triumphs all seemed to be marked by Amy's ability to connect with how her body felt instead of how it looked. One day, a couple of years after she finished college, Amy passed a store that had bins of sale items outside. She noticed a big bin of 100 percent cotton women's underwear. It stopped her in her tracks. It was a strange place for a turning point, but you just never know when key moments are going to arrive. It ends up that even though she was an adult, Amy had never really bought her own underwear. Her mother always bought her underwear, what Amy described as "little tiny panties, always the wrong size and always uncomfortable." Amy was on her way to a meeting at the time. She walked into the store, purchased a pair of the cotton panties, and put them on in the restroom before her meeting. She remembers that moment with obvious joy. "I put on my first pair of big-girl cotton granny panties!"

"How did it feel?" I asked, oddly happy for her.

"So great! So comfortable. And I realized I was a little bit cranky because I was always a little uncomfortable. My clothes were always a little too small because I wanted to be in the smallest size I could possibly be in or I wanted stuff that was tight so it made me look smaller. I put on this pair of big old black granny panties and it was so great. I felt so comfortable." Amy cracks up as she remembers that day.

"Did you keep wearing them?" I asked.

"Oh hell yeah," Amy replied. "Hell yes, I did. And it really was this act of kindness for myself. It was this act of acceptance, like, I am not going to force my body into clothing that is not made for me."

It's hard to appreciate our bodies when we live in a culture that tells us they are never good enough. But remembering what our bodies do for us is an essential first step in finding the ability to feel at home in our own skin. Though I've focused a lot on physical activity, remember that thinking about your body in terms of what it can do is not just about working out. You don't have to be fully able-bodied to feel gratitude for what your body can do. You don't have to run a marathon or complete CrossFit challenges to be worthy. That's not all there is to your body's function. Your body is home to all of the skills you've developed over your lifetime. It facilitates your important social interactions. The movements of your face express your deepest emotions. And the internal functions of your body, those you can't readily see, are just as inspiring. Your body takes nutrients from your food and uses those nutrients to power you as you make your way around the world. How can a body that does all those things be disgusting or shameful? It's the chorus of objectifying voices in our culture that blinds us to these wonders.

In one study my lab conducted, we asked women to write a letter to their body, expressing gratitude for everything their body does for them. Their responses were jubilant tours through the myriad bodily functions that keep us going every day. Here's an excerpt from one of my favorite letters from this study, written by a nineteen-year-old woman.

Dear Body,

Thank you for letting me accomplish all that I have sought to do in life. You allow me to get myself out of bed every morning before I'm even fully aware. Thank you, legs, for letting me walk to class when I'm on time and run when I'm not. Without you, I could have never danced, played basketball, or gone hiking. I'd never know what it was like to run like a little kid without a care in the world.

Thank you, hands, for letting me paint, write, and type this letter out right now. Without you, I'd never know what it's like to hold the hand of someone I love. Thank you, eyes, for letting me see the world, the good and the bad. Without you, I'd never see the faces of my parents, my siblings, or my friends. I'd never see a sunrise, or a rainbow, or any of the sights I've been lucky enough to witness.

Without my body, I'd never know the feeling of falling into bed at the end of a long day, or lying on the beach under the sun. I'd never be able to experience my mom's hugs.

I'm sorry if I take you for granted sometimes, because you're pretty amazing. I'll try my best to take care of you, and appreciate everything you let me do every day.

While we can turn down our focus on appearance, we should not and cannot ignore our bodies. Instead, we need a different way to think about our bodies, one that leads to greater health and less beauty sickness. Thinking about your body in terms of all the things it can do is healing and empowering. It combats body shame and body dissatisfaction. When we remember that our bodies are for doing things, we will be more motivated to care for them—no matter how they look.

SOMETIMES IT'S HARD FOR AMY to imagine that she ever could have been the high school girl who cried after completing an obstacle course in front of her classmates.

"Flash-forward ten years," she says, "And I'm doing this twelve-mile obstacle-course run called the Tough Mudder for fun. And I am having these moments when I'm climbing over things and being lifted and hoisted and jumping."

"And is there any shame?" I ask. "Any worry in those moments?"

"No," Amy responds with confidence. "It's just the sheer joy in doing it, and of doing it with my friends. If you had told my sixteen-year-old self that one day I would be running an obstacle course for fun . . ."

"Your sixteen-year-old self would be shocked?" I asked.

"She would have laughed you out of the room! It would have been inconceivable to me that this is something that I could do for pleasure." Amy told me about an obstacle, a twenty-foot-high wall that is slanted toward those climbing it. It's called Everest.

"How are you supposed to get over that wall?" I asked.

Amy grinned and leaned forward. "The idea is you take a running start and you run as far as you can up the wall, and then grab the lip and hoist yourself up. And people wait up at the top and dangle ropes down and dangle their arms down, but still, it is a really hard obstacle and most people don't do it. And the first time I did the Tough Mudder, my friends went before me and were leaning over and I just ran. I ran up that thing and I leaped up and I grabbed someone's arms and they yanked me up and I made it to the top. And it was a feeling that I just cannot describe. I was screaming at the top of that thing and pumping my fists, *Top Gun* style. It was something to know that I could do it. And I did it fat."

"You didn't have to be someone else."

"I didn't have to be someone else to do it," Amy confirmed. "I didn't have to be the ideal. I didn't have to be what was expected of me. And I am never going to be that, because I don't want to be. Because the price I would have to pay to be just skinny is too high. There would be nothing left of my life."

Amy has a young niece who is just entering her teen years. I asked Amy what her hopes are for that niece in terms of her relationship with her body.

"I hope that she finds something physical that she likes to do so that she can feel that joy of engaging with your body in a way that makes you feel strong. I mostly hope that she realizes that she is fantastic at whatever weight she ends up being. But the truth is, I don't even know if that is possible."

I understand what she's saying. Given the world we live in, it sounds a bit pie in the sky, this idea that you can accept your body no matter what. Amy sighed a bit. "I think the best I can hope for is that she gets to a place where she feels that way sometimes. That she feels that way often enough that she remembers what it feels like when she doesn't feel that way."

"I think that's a beautiful thing to hope for her," I responded.

"That's what I hope for myself too," Amy said.

14

Learning to Love Your Body and Teaching Others to Do the Same

A FEW YEARS ago, one of my research assistants asked me why we weren't studying women who felt good about their bodies. Our lab studied body image, but why did it always have to be negative body image? She had an excellent point. We have a lot to learn from women who manage to swim upstream, finding ways to stay healthy in this sick culture. My lab set out to find some of those young women and invite them in for interviews. We devised a straightforward set of criteria. We wanted to talk to women who had low levels of body dissatisfaction and who weren't engaging in eating-disordered behaviors. The plan sounded simple, but our search didn't work out quite like we'd imagined.

We screened hundreds of college women, but never really found what we were looking for. Sometimes a woman would seem healthy based on her survey scores, but then we'd bring her in to our lab for an interview and she'd say, "No, I don't really like my body. I just know it's not good to feel that way, so I didn't admit it on the questionnaire." Other women told us they loved how they looked, but then admitted to starving themselves on a regular basis to get there.

When I look back at that research endeavor, I realize we were going about our hunt all wrong. Instead of searching for women who felt good about their bodies, we looked for women who didn't feel bad. There's a real problem with that approach. Not hating your body is different from actively appreciating your body. And being free from eating disorders doesn't necessarily mean you take good care of your body. Think of it this way: Just because you're not depressed on any given day doesn't mean you're happy.

There was another problem with our approach as well. We hadn't considered that it's possible to appreciate and love your body overall, but simultaneously feel dissatisfied with some aspects of it—just as you can be a generally happy person, even if you feel depressed at times. Similarly to how my lab had spent a lot of time documenting the problem of beauty sickness without coming up with too many solutions, we'd also failed to take a good look at the women who had the most to teach us about overcoming beauty sickness. We wanted women to stop hating their bodies, but hadn't thought as much about what it would mean for women to truly enjoy their bodies and care for them without too much struggle.

The distinction I'm talking about here is between positive and negative body image. It's always so much easier to focus on the negative. It's more visible, more worrying. In some ways, it's also more straightforward. But if we want to understand how to combat beauty sickness, we need to think of the flip side of the coin. We need to consider how some women learn to have a truly positive body image and how we can successfully teach those healthy attitudes and behaviors to others. If we want to create a better world for girls and women, we can't just turn down the volume on the negative; we also need to turn up the volume on the positive.

Let's start by meeting Hannah*, a white, twenty-nine-year-old high school teacher in Denver. Like most women, Hannah has her share of days when she finds herself dissatisfied with the image in the mirror. But thanks to having been raised to think of her body, food, and exercise in healthy ways, Hannah has more good days than bad. We can learn a lot from the subtleties of how she thinks about her body.

Hannah has lived all over the place. She went to college in New England, then lived in Jerusalem for a year afterward, teaching English to fifth-graders. Her first teaching job in the United States was in Mississippi, which she followed with two years in the Midwest to earn a master's degree. Hannah has now settled in Denver with her fiancé, where she enjoys the sunny weather and outdoor activities.

Hannah and I met for our interview at a Dunkin' Donuts one late afternoon. I drank a giant iced tea while we sat at a little table in the mostly empty shop and talked through the life course of Hannah's relationship with her body. I started by asking Hannah to share the earliest lessons she learned about appearance. That question quickly brought us to the topic of hair.

"As you can see," Hannah said with a smile and a flourish of her hand, "my hair is very curly." She pulled one springy curl down and let it bounce back up.

Hannah told me that when she was young, she had no idea what to do with her mess of curls. Hannah's mom had curls too, but she wore her hair so short that Hannah figures she never thought much about them. Young Hannah attempted to straighten her hair, but the results weren't so great. One day, her seventh-grade teacher pulled Hannah aside and said, "You have curly hair. Here's what you need to do."

"She tutored you in curly hair?" I asked.

"Yes!" Hannah smiled. "She gave me advice. I would come in and show her. Then she'd say, 'That looks good. Now tomorrow, see if you have some gel or mousse, try that and see how that goes.'" This process went on for a week or so, until Hannah got a handle on her curls.

Hannah remembers these interactions fondly, but I had a hard time getting on the same page as her. Something seemed off to me about a junior high teacher being that concerned about how Hannah's hair looked. Why should a girl in school need to worry about what her teacher thinks of her hair?

But Hannah explained that I had it all wrong. She attended a very small, intimate Jewish day school. Students were close with each other and with their teachers. She said the teacher had curly hair herself and could probably tell Hannah was spending hours trying to straighten her own hair every day. So Hannah saw the teacher's intervention as an act of kindness. She was trying to teach Hannah to be comfortable with the hair she had. It ended up being a powerful gift, saving Hannah time and energy and all sorts of needless struggle. Today Hannah spends very little time on her hair—she says that's the thing she likes most about it, that it requires so little attention.

Being comfortable with her curls wasn't always easy for Hannah. She remembers watching all manner of shampoo commercials when she was young, and it seemed they all made reference to "silky" hair. Hannah learned that straight hair, especially blond straight hair, was what our culture found most beautiful on women.

"My hair is never going to be silky," Hannah explained to me. "I would think about those women with hair that would just fall." Hannah makes a swooshing motion as she says this, from the top of her head to her shoulders. "In the commercials they would toss their hair around. It just looked so soft. And I always liked the way messy buns looked with straight hair."

For a while, even after her teacher instructed her on managing her curls, Hannah would still straighten her hair for special occasions. She and I chuckled about how women with curly hair straighten it when they want to look extra good, and women like me—with stick-straight hair—do the opposite and try to curl it on special occasions. I have about as much luck curling my straight hair as Hannah did straightening her curls. We both agreed that there's something strange about a culture that requires you to look unlike yourself when you want to look good.

Hannah's getting married in a few months. She's on a quest to find a stylist who knows how to deal with naturally curly hair. She wants to look her best for the wedding, of course. But it's an act of self-love that she also wants to look like herself. She doesn't want her fiancé to walk down the aisle and see a stranger. She doesn't want to have to "perform" the role of bride.

Hannah hasn't straightened her hair in over seven years. She says, "I don't want it straight now. It looks weird. It doesn't look like me.

"Now," she says, "I like my curly hair a lot." Hannah pats it with a smile.

Part of what is striking about how Hannah talks about her looks is the kindness you hear in her words. It's not that she gazes in the mirror and adores every feature she sees. She wishes she could lose about five pounds, for example. But Hannah seems to hold no malice toward her reflection. She talks about her body with a gentle acceptance that I rarely hear from young women.

Growing up, Hannah was aware that the beauty ideal in media images was at odds with elements of her own appearance, but she was held in a protective cocoon that kept those images from doing too much damage. Hannah was part of a tight-knit family, with four daughters all relatively close in age. She was surrounded by girls and women who looked much like she did. As she put it, "Tan, thin, and blond was in the media, but not something that I saw at all in my real everyday life." During moments of doubt about her appearance, Hannah couldn't be too down on herself without inadvertently questioning the appearance of her similar-looking sisters and her mother as well. She loved the women in her life and wouldn't dream of criticizing how they looked. This care for others seems to have played a role in her self-care.

Hannah has thought a lot about how so many girls and women struggle with body image, eating, and appearance concerns. She hears about this pain regularly from the adolescent girls she teaches. Though she told me she and her sisters probably all had brief periods where they worried about something related to body image, Hannah says all four women ended up "with a healthy dose of self-esteem." When I asked Hannah what the source of that self-esteem was, she gave her mom a substantial portion of the credit.

For the past few years, researchers have started to more fully explore what it means to have a healthy relationship with your body. Positive body image is multifaceted, marked by a variety of attitudes and behaviors. First, women with a positive body image tend to think about their bodies in terms of what they can do. Just like Amy, the hiking, bike-riding, wall-climbing woman you met in the previous chapter, these women foster what's called a functional orientation toward their body. They recognize all the tasks their body accomplishes and respond with gratitude. As a result, they see their body as something to take good care of, not something to beat into submission through constant diets, punishing workout routines, or cruel words.

At the Table Together

Hannah was never taught, as so many women are, to see food as the enemy. She explained, "My mom was a really good model. She's never been super thin, but she modeled for us what it's like to do healthy eating, but also splurge occasionally on things we like. I can remember her saying multiple times over the course of life, 'I'm just going to enjoy this meal with my family. I'm not gonna worry about counting calories.' And she wouldn't binge either, but she was just, like, 'I'm gonna enjoy this. That's more worth it to me than stressing about what I'm eating.'"

"So you grew up learning that you're allowed to get pleasure from food?" I asked.

"Yes. Right," Hannah confirmed. "I definitely credit my mom for a lot of that, for teaching me a healthy relationship with food—and I'm a big food person. Based on my upbringing, food is about bringing people together, creating community and hospitality. I host people. I love to cook."

"It's a way of showing love for people?" I asked.

"Yeah. Definitely," Hannah agreed. "And both of my grandmas, that was how they communicated love—through cooking, baking."

Hannah may not have realized it, but she was making an important point. Women who engage in restrictive diets don't just face the emotional burden of chronic hunger. They also risk losing touch with the community building and social connections that are so often fostered through shared meals. Dieting can be a recipe for loneliness and isolation.

Women with a positive body image have the ability to enjoy food without overstuffing themselves or engaging in crash diets to "make up for" what they've eaten. When a group of researchers conducted in-depth interviews with women identified as having positive body image, they found that these women were experts in the practice of *intuitive eating*.[1] Intuitive eating involves listening to your body for hunger cues and using those cues to determine what and how much you eat, instead of eating to manage emotions or eating something simply because other people around you are doing so. There's an element of mindfulness to intuitive eating. It requires attention to and respect for your body's processes.

Though she never used the term, Hannah strikes me as a practitioner of intuitive eating. She's sensitive to her body, listening to its signals. Hannah said she's never weighed herself. I struggled to reconcile this statement with her earlier claim that she'd like to lose five pounds of recently gained weight.

"How do you know you've gained five pounds if you never weigh yourself?" I asked.

"Because I know my body," Hannah responded. "I know how it feels, how it fits into my clothes. I don't count calories. I don't weigh myself. That's something my family doesn't do, so I didn't see it growing up and now I just don't do it. We've never been like that."

I asked Hannah how, if she doesn't count calories, she's been trying to get back to her typical body weight. She explained, "I know what's healthy and what's not. Like, I don't need to know how many calories are in a donut. I know it's not particularly healthy." I laughed at her example, looking at the wall of donuts to my left.

Families in Front of the Mirror Together

Although girls' attitudes toward body image, eating, and exercise are shaped by numerous factors, mothers can often play an outsized role. When I hear young women engage in particularly vicious bouts of negative body talk, I'll often ask where they learned to talk that way. These women never tell me they learned the practice from movies or television or their friends. They tell me they learned it from listening to their own mothers talk that way about themselves.

One study my lab conducted found that regardless of their own body size, women who believed their mothers were body dissatisfied were more likely to be body dissatisfied themselves. When mothers disparage their looks in front of their daughters, they teach those daughters that body hatred is not just acceptable but part of being a woman.

In a recent study in the UK and Australia of over 200 mothers and their five- to eight-year-old daughters, researchers found that the more mothers self-objectified, the more likely their young daughters were to be engaged in a beauty culture typically associated with teens and women.[2] These very young girls had already picked up on the lesson that being female means both performing beauty for those around you and being dissatisfied with your own reflection. The daughters of mothers who self-objectified were more likely to already be wearing makeup and high heels, to comment on their own or others' appearance, and to worry about how they look in photographs. They weren't even teens yet, and they were already bearing the burden of beauty sickness.

Well-known author Jennifer Weiner recently penned an op-ed for the *New York Times* in which she described the friction of wanting to raise daughters who aren't focused on appearance, but feeling dissonance over doing so when she has yet to get to that place herself. She asked, "How do you preach the gospel of body positivity when you're breathless from your Spanx? How can you tell your girls that inner beauty matters when you're texting them the message from your aesthetician's chair?"

Hannah's mother did something very important in terms of fostering positive body image among her daughters. She didn't just show acceptance toward her own body, she also refrained from commenting on her daughters' weight.

At one point, Hannah's youngest sister, Alyssa, went through what Hannah described as "a chunky phase." People kept asking Hannah's mom when she was going to "talk to Alyssa about her weight." Her mom's response was "Never! Unless *she* brings it up. Then we'll talk about it."

Hannah really respected that decision. As she remembers it, "My sister was such a happy kid, just the happiest kid. Why would my mom bring that up? Why would my mom do that?"

Hannah's instincts about her mom's decision are consistent with research on the topic of parents' and children's interactions around weight. While it's a great idea to teach your children about healthy eating and exercise, focusing on weight tends to backfire. It's more likely to lead to hurt feelings than healthy habits. A recent study out of Cornell found that women whose parents commented on their weight when they were young were much less likely to be satisfied with their bodies as adults, regardless of the actual size of their bodies.[3] They didn't learn to be thinner; they just learned to be unhappy with what they saw in the mirror.

Negative comments by parents about the appearance of other women likely play a similar role. These types of comments suggest to both sons and daughters that it's acceptable to judge women based on their appearance and denigrate those whose appearance doesn't fit the ideal. When fathers are openly critical of other women's bodies, that teaches their daughters a particularly potent lesson. It suggests that being loved by a man is contingent on being beautiful, and that women exist to look good for men.

When it comes to mothers and daughters, behaviors likely matter even more than words. Daughters are quick to observe their mothers' actions and emulate them, even when doing so is inappropriate, unhealthy, or even dangerous. A study out of Penn State and Washington University found that regardless of their weight, girls were much more likely to start dieting before age eleven if their mothers were dieting.[4]

This finding is more serious than you might realize. Early childhood dieting is linked with depression and disordered eating, and contrary to what many would guess, it predicts a greater likelihood of being obese as an adult. This does not seem to be because those prone to weight gain start dieting early. A fascinating recent study out of Finland examined pairs of identical twins to test whether dieting could actually *cause* weight gain later in life.[5] The researchers focused on sets of twins that were "discordant" for weight-loss attempts. In other words, one twin had a history of dieting and the other did not. They found that over time, even though these pairs of twins had identical genes, twins who had a history of dieting were more likely to gain weight than their nondieting counterparts.

Just as mothers can teach young girls to dislike their bodies, they can also teach them to love and appreciate their bodies. In an Arizona State University study of 151 mothers and their five- to seven-year-old daughters, girls and their mothers were first separated and taken to different rooms.[6] Researchers asked them to stand in front of a full-length mirror and look at their whole body, from their head to their feet. Then they instructed the mothers and daughters to talk out loud about what they liked or disliked about their different body parts, including their hair, face, legs, and stomach. After completing this mirror task individually, the mothers and daughters were reunited in front of a mirror and asked to complete the task a second time together. The moms were always given the opportunity to remark on their body parts first, before their daughters followed suit.

The daughters in this study regularly changed their initial responses to be congruent with their mothers' responses. Girls who had previously said positive things about their stomach, for example, switched to saying something negative after hearing their mom disparage her own stomach. But more important, mothers who said positive things about their own bodies prompted their daughters to do the same.

I asked Hannah if she thought her mom approached her daughters with a deliberate plan to build positivity, or if perhaps it was more of a natural extension of who she was. Hannah thought about it for a moment, then responded. "I think it was more the latter, an extension of who she was, but it was a little bit intentional. She is aware of all the pressures and expectations, and she wanted us to feel good in our own skin, in our own bodies."

Hannah tells a delightful story of a positive body image moment that's become lore in her family. Alyssa, Hannah's little sister, came home from school upset one day. She felt like all of her classmates were skinny girls, and that she was the only one with a different body type. She cried to her mom, saying, "I'm just not like the other girls." Alyssa was normally such a happy person—her distress seemed worse for the fact that it was so out of character.

Hannah explained, "She was just so sad. And my mom said, 'Well, Alyssa, there must be one thing you like about yourself. What's one thing you like about yourself?'"

Young Alyssa responded earnestly, "I like my whole face!"

Hannah's mom beamed. "You like your whole face? I bet those other girls couldn't say they like their whole face, just as it is!"

Hannah told me she and her sisters will still say, "I like my whole face!" and smile at the inside joke.

I asked Hannah, "Do you think Alyssa still likes her whole face?"

"Yeah! I think so," Hannah responded, nodding happily.

The Benefits of Body Appreciation

Hannah describes her mom today as someone with the ability to appreciate her own body. "I think she would say she feels good in her own skin," Hannah says. "I think she still sometimes struggles, and I think as you get older it's harder to maintain weight and exercise becomes harder because everything hurts, things like that. But I think she's an attractive person and feels good in her body and takes pride in her appearance."

Hannah found her mom beautiful growing up, even if she never had the lithe body of a fashion model or the silky long tresses seen in shampoo commercials. Although Hannah never received direct messages from her dad about beauty, she watched how her dad treated her mom. "He expresses certain things with touch more than words," she explained. "He was affectionate with her, and I would see that, that positive feedback—even if it wasn't explicit and verbal."

"It must have been nice to see that growing up," I said.

"It was. It was," Hannah responded.

It's not that women with a positive body image don't care about how they look. They still enjoy the moments when they feel attractive. They just don't hinge most or all of their sense of self-worth on whether others find them attractive. If these women do things like put on makeup or get their hair done, they see these behaviors as a way of caring for themselves instead of a means to perform the role of "beautiful woman" to the outside world. Their fashion choices are about comfort and self-expression, not the need to create desire or jealousy in others.

They love their bodies, but not out of a sense of narcissism or vanity. Their appreciation for their bodies is not contingent on others finding them sexy. They avoid engaging in negative body talk, actively role modeling body appreciation to other girls and women. They respect their bodies even when they differ from the appearance ideals we see in media images. They focus on the strengths of their bodies and take the flaws with a grain of salt.

Researchers out of the Ohio State University identify "body appreciation" as a key element of positive body image.[7] To assess women's body appreciation, they ask women to rate their agreement with survey questions like "I respect my body" and "Despite its imperfections, I still like my body." In studies of hundreds of women, they found that this ability to appreciate your body predicts myriad positive outcomes. It's correlated with greater self-esteem, increased optimism, and more frequent use of healthy, proactive coping styles. Body appreciation is also associated with less body dissatisfaction and less disordered eating. A recent survey of around 250 women in Australia found that women with more body appreciation were more likely to take care of their body in a variety of concrete ways. They used sun protection more often, were more likely to screen their skin for cancer, and were more comfortable seeking medical attention when they felt they needed it.[8]

Body appreciation can also act as an important buffer, protecting women from the onslaught of reminders that their bodies don't match the cultural ideal. In interviews with women identified as having unusually positive body images, researchers found that these women learned to block out messages that could challenge their ability to appreciate their bodies.[9] They were certainly aware of those messages, but they worked to create a filter to lessen their impact. They focused on letting positive body messages in and letting the negative messages bounce off. In a different study, over 100 women in England were shown ultrathin idealized media images of women. Those who scored high on body appreciation seemed to be protected from the negative effects of exposure to these images.[10]

As women age, their bodies tend to move farther from the cultural ideal. Yet researchers have consistently found that body dissatisfaction tends to stay relatively steady as women age. A key explanation for this finding is that as women age, they learn to appreciate their bodies more for what they can do. Even if they look less like women on the covers of magazines, they've had more years to be amazed by all the things their bodies have done.

Paradoxically, feeling the loss of body functionality with age can also put women in the mind-set of appreciating the functions they still have. In one study of nearly 2,000 women aged fifty and up,[11] researchers found that although many of these women still felt bound by cultural pressures to be thin and traditionally attractive, others found a freedom with age, one that was fueled by focusing on their bodies' functionality. They were appreciative of their bodies even while recognizing weaknesses and flaws. Some of these women expressed sorrow about having spent their younger years feeling ugly or dissatisfied with their bodies. It seemed like such a waste of time, looking back.

Hannah works to incorporate appreciation into her own experience of her body. She explained, "I don't love my body all the time. But when I look at my body in front of the mirror, I try to think about how I do yoga. I try to be grateful for this body that works. I try to think of my body as a body that works rather than a body that's beautiful. Sometimes that makes me feel better when I'm feeling not so good on a given day. It's like I'm blessed with a body that I can do stuff in. It's bringing myself back to that as opposed to a pretty body."

Hannah has faced the struggle to continue appreciating her body as it changes with age. She started getting gray hair around age twenty-three, and described this as something she sometimes fixates on. I looked at Hannah as she said this and admitted I couldn't see one gray hair on her head. But Hannah assured me they were there. Part of Hannah's struggle with her graying hair is that she was utterly unprepared to see it at such a young age. She thought women got gray hair in their fifties. It ends up that as much as many of the women in Hannah's life are positive role models for body acceptance, they drew a line when it came to gray hair. Hannah had no idea that they'd all been dying their hair for decades. That's why her first few grays came as such a shock.

"Will you dye your hair?" I asked.

"I don't know," Hannah responded, looking somewhat anguished. "Because I really think, in general, I'm the kind of person who tries to look at my values before I act. In this situation, my value would be not to dye it, but there's part of me that would want to."

"What values make you not want to dye your hair?" I asked.

"I think of being comfortable with aging and not feeling like I need to do something that everyone else does. It's also expensive, and I guess in terms of values, I don't know if that's where I want to be spending my money every month."

Hannah remembers a sweet story. One of her cousins recently caught her looking at her gray hairs in the mirror. Hannah explained, "He had this teacher in high school who was brilliant, wonderful. Just had, like, a really good aura about her. Dr. Klein. And he said, 'Hannah! Stop worrying about your gray hair. You're gonna look just like Dr. Klein!'"

"What do you think about looking like Dr. Klein?" I asked.

"She's got salt-and-pepper hair. It's actually not bad. It's a nice look. And it was sweet, because it was like, this is a woman that he so admires, and he's, like, 'You'll look like Dr. Klein. It'll be great!'" Hannah grins at the memory.

Pretty Amazing

Even if parents work hard to create a home filled with body appreciation, it isn't easy to push back against the power of beauty sickness. Well-meaning parents still have to contend with a barrage of media influence. On top of that, beauty-related social reinforcement outside the home starts young, when girls are first noticed and praised for being pretty over anything else they might do, say, or be.

Many parents' instincts tell them to build their daughters up by telling them they are beautiful. But as we've already discussed, compliments about appearance don't actually seem to make girls and women feel better about how they look. Instead, they're just reminders that looks matter. Poet Rupi Kaur captured this sentiment perfectly when she wrote:

> *i want to apologize to all the women*
> *i have called pretty.*
> *before i've called them intelligent or brave.*
> *i am sorry i made it sound as though*
> *something as simple as what you're born with*
> *is the most you have to be proud of*

In one study my lab conducted, we asked college women if they could recall a time when they complained to their mother about how the women they saw on television were all so much thinner or more beautiful than they were. If the women remembered that type of interaction with their moms (and most did), we asked them to tell us more about it. First, they shared how their moms actually responded in that scenario. Next, we asked them how they *wished* their mothers would have responded. Then we looked at the differences.

Most of the women in our study said their moms would respond to their distress by telling them they are beautiful just the way they are. The young women we studied didn't necessarily think this was a terrible response, but they told us they wished their moms had taken a different approach. They wanted their mothers to emphasize how important things beyond beauty are.

When we asked how these young women would respond if in the future they had a daughter who faced a similar struggle, they wrote things like "I would remind my daughter of some of the great things in her life that couldn't be the same if she only concentrated on her body." Another wrote that she would tell her daughter that "In order to be truly attractive, you must accept and take care of your body and cherish and nourish your mind. You must care about other people and strive for knowledge. A pretty body is not everything."

Both mothers and fathers have the opportunity to create homes that provide an antidote to the broader cultural forces that seek to shape their daughters. Girls get enough messages out there in the world that their worth is determined by their appearance. The challenge for parents is to foster a climate that focuses on things other than how people look. Girls are helped not by compliments about their own appearance, but by a steady ongoing focus on things that matter more than appearance. If you want to compliment a girl or woman, compliment her on something she can actually control. Reinforce the idea that being hardworking, focused, kind, creative, and generous matter. None of these qualities require any particular body shape or hairstyle. Tell her you notice how much effort she puts into the things she cares about. Tell her that you enjoy spending time with her because she is interesting. Tell her that she inspires you and then explain why or how.

When I talked to Hannah, she told me that she hoped to have children one day. I asked her to fast-forward to a time when she had a twelve-year-old daughter and to imagine that her daughter came home from school saying, "Mom, I'm so fat and ugly."

"What would you say to your daughter?" I asked.

Hannah looked down at the table, stricken by the idea that her own daughter might suffer in that way. She responded slowly, "I would first hug her, and I would talk to her about it."

She thought back to a class she took when a professor told her, "The best thing you do for your kids is not to say, 'No, you're not fat and ugly,' because that denies what they feel, so don't say that." Instead, Hannah said, she would "talk about what they like about themselves, talk about what I love about them."

The question triggered a memory in Hannah, one from when she was a teenager away at summer camp. One of Hannah's friends at camp, Abby*, was a girl everyone considered "hot."

Hannah recalled, "I remember Abby saying to me once, 'It's not always fun being really attractive, because you never know if someone likes you for you.'" Hannah laughs now, thinking back to her reaction at the time. "I was just, like, 'Ugh, Abby.'" From Hannah's perspective, everything seemed so easy for Abby.

"Did you believe what she said at all? That sometimes it's hard to be really attractive?" I asked.

"No!" Hannah responded. "I think it was a totally new concept to me, that someone might not see all the parts of her because they were so caught up in how she looked. I never would have thought of that." Now Hannah wonders if Abby might have had a point.

I asked Hannah if she felt that people saw all the parts of her. Hannah thinks they did. She was often chosen as a confidante, because people recognized her as loyal, trustworthy, and understanding. She describes those qualities as a core part of who she is. Hannah eventually became a counselor at the same camp she attended as a child. She spent five summers doing that work. She said, "Boatloads of girls would come to me and be, like, 'Is anyone ever going to notice me? Is anyone ever going to love me?'"

"What would you say to them?" I asked.

Hannah was enthusiastic in her response. "I'd be, like, 'You're awesome!'" She cheered. "They were these awesome, smart girls. And it was so sad, it was so heartbreaking to see that. I would listen to them and pump them up, and point out all the qualities that I thought were wonderful about them." Basically, Hannah described an approach many researchers would have recommended.

She continued, "I would be, like, 'Look at all of you! Look at all the pieces of you!' I would tell them to be themselves, because as you get older, what's most attractive to someone else is your own comfort in your own skin, both your body and your identity."

Compassion as a Tool for Building Body Acceptance

It's easy to talk about appreciating your body in the abstract, but in dark moments of self-doubt, it can be hard to get there. Self-compassion, a practice that psychologists have recently begun incorporating into treatments for anxiety and depression, can be a good starting place for fostering a more positive view of your body. Self-compassion involves treating yourself with warmth and kindness, and accepting that part of being a human is having flaws and imperfections.[12] It's a way of connecting to others, acknowledging that we are all flawed.

Women who practice self-compassion can more readily appreciate the diversity of human bodies and the uniqueness of their own features. Unlike self-esteem, which generally depends on the evaluations of others, self-compassion doesn't require approval or input from other people. You don't need any special tools to offer compassion to yourself. It only takes practice. Hannah modeled this type of compassion when she talked about how she hoped to interact with her own children one day.

Hannah has been on the receiving end of recommendations to get a nose job on several occasions, but she's always held out and is glad she's done so. I asked her how she would respond if one day she had a daughter who came home and said, "Mom, I really want a nose job. I hate my nose."

Hannah said she would say, "I used to think about that too! But my nose is part of who I am, and I'm okay with that. There are parts of ourselves that we like more than others, and that's okay." One of my research assistants recently led a series of studies testing whether this type of self-compassion could change the way women feel about their appearance.

We brought several hundred women into our lab and randomly assigned them to write one of several different types of letters to themselves. They wrote either a self-compassionate letter to their body, a letter focused on their body's functionality, a neutral letter describing themselves, or a neutral letter describing their body. You read one of the lovely function-focused letters in chapter 13. After they got their assignment, we left them alone at a computer for ten minutes to write their letter. Then we asked them to spend an additional five minutes reviewing and editing their work. What we were most interested in was the potential impact of writing a self-compassionate letter to one's own body. We based our instructions for these letters on exercises available at researcher Kristin Neff's site, self-compassion.org.

We instructed the women in our self-compassion condition to

write a letter about their body from the perspective of "an uncon-ditionally loving imaginary friend." We indicated they should im-agine that friend is aware of any flaws in their body, but also their body's strengths. We asked them to write what that friend might say to show their body is appreciated just the way it is.

The letters our research participants wrote were inspiring. When we read them together in a lab meeting, they took our breath away. Here's one of my favorites:

> Well, friend, here goes.
>
> You listen to the images, the media, the voices that tell you that your body is wrong, undesirable, and flawed. But listen to me.
>
> Your body is amazing. You find imperfection, but I find power. I see muscles that work just the way they are designed to; it couldn't matter less whether they are visibly bulky or toned. I see a stom-ach that heaves with laughter and arms that comfort. I see soft shoulders for leaning on and elegant fingers for making music. I see eyes of understanding and a smile that can light up a room. I see skin that is kissed but not burned by the sun, though in any shade it glows. Curves are not ugly, but stunning and useful. There is no need to punish, only to care for this vessel of your soul.

For the women in this study, all it took was a brief paragraph of instructions to open floodgates of body compassion. Another woman wrote:

> Dear Me,
>
> I know you struggle with the way you look, you always have. (And let's be honest, who hasn't?) The mirror and scale have be-come your closest frenemies ever since you were in the 7th grade. You nitpick over the tiniest of rolls, the onset of dark circles, and

that weird wart thing on your foot. But none of those things define you, none of those things are actually worth the space they take up in your head. Your body gets you through life, it's that one everlasting vehicle that steers you through your college degree, your marriage to perhaps the greatest boy you've ever been with . . . the possibilities are truly endless. Your body—whether emaciated, slim, curvaceous, or pudgy—will get you through it all. It won't discriminate. All the hardships you've endured, all the pain you've emotionally and physically gone through because of hating your skin and extra flab were unnecessary for your mental health. When I look into your eyes, I see vitality. I see someone in there who is more than just her mortal coil. She is a lover, and deserves to see herself that way. Everything on this planet will either fade away or die; you are no exception. But the most beautiful thing about existence is that you have this finite time to create and love, so why waste precious time beating yourself up when you could simply embrace the infinite light that shines out of every pore on your skin? I'm telling you, that's the only way. When you look past the shell that contains your soul, you begin to view others the same. Less judgment, more acceptance. More love.

Another woman wrote openly of a painful struggle.

I know that you used to self-harm because you didn't feel beautiful enough and I remember the hours you spent upset about not fitting society's idea of beauty. If you don't get anything else from this letter, please remember that I will always love you. No matter what you do, say, become, nothing will ever change that and I am always here to unconditionally support and care about you. So you may not be the thinnest or have the clearest skin, but at least you're you. All of these things, these imperfections and flaws, they're part of what makes you you. I'd rather you be who you want, feel, and believe you should be rather than a copy of what society dictates. Don't be afraid to love yourself and receive that same unconditional love from others.

Love,
Someone who cares very deeply about you

The letters were all so beautiful that it's difficult to stop there. I could fill a book with the words these women wrote to their own bodies. But what's even more important than what they wrote is the impact their words had. Compared to those in the control conditions, the women who wrote these self-compassionate letters ended up with significantly higher body satisfaction and a more positive mood.

Other research has come to similar conclusions. In an Australian study of over 200 women aged eighteen to thirty, researchers found that greater self-compassion was linked with more body appreciation.[13] Self-compassion was also associated with less consumption of appearance-focused media, less self-objectification, and fewer appearance-related social comparisons.

We can and should work within our own communities of loved ones to eradicate beauty sickness, because it's wonderful when positive body image comes from receiving accepting messages from those we love. But even when we can't convince our loved ones to take this more positive approach, we can still choose to offer our own bodies compassion. Self-compassion is the perfect tool for replacing the negative body talk we so often hear. If you must talk about your body, talk about it with this type of kindness and understanding.

When you show yourself compassion, you're also modeling the

compassion you'd like others to show. Not only are you helping yourself, but you're helping countless others who might see your example and follow it. The more you practice compassion toward your own body, the more easily you will be able to turn away from the mirror and toward the world.

15

Turning Away from the Mirror to Face the World

THERE'S A BURDEN to having to perform when you should be just living your life. It changes you. It directs your energy to monitoring others' reactions instead of your own feelings, needs, and desires. But the more at home you feel in a given setting, the less you should feel like you have to perform and the more you should feel free to just live.

Beauty sickness is problematic, in part, because it stops us from feeling at home in our own bodies. It causes us to see our bodies as something for other people instead of for us. The more we can feel at home in our own bodies, the less we will feel we must constantly perform "prettiness" to the world. When we stop performing, we free up mental resources for other tasks.

The stakes here are real, and they're significant. We cannot live our lives fully when our appearance is constantly under evaluation. We cannot make progress toward our goals if we believe that all possible success and happiness hinges on the result of that evaluation. That sense of having our bodies constantly appraised creates a mirror in our minds. It's looking in that mirror too often that makes us beauty sick.

Beauty sickness matters in part because it hurts. But even more important, it matters because it's hard to change the world when you're so busy trying to change your body, your skin, your hair, and your clothes. It's difficult to engage with the state of the economy, the state of politics, or the state of our education system if you're too busy worrying about the state of your muffin top, the state of your cellulite, or the state of your makeup. There is work to be done in this world. Leaving the world in better shape than how you found it is more important than the shape of your body.

The women I know are all in when it comes to issues they care about. They want to create safer, healthier, more just, and more vibrant communities for themselves and others. Beauty sickness makes it harder to reach for those goals. Before we can *lean in*, we need to step away from the mirror. You cannot command the boardroom if you're distracted by worries about the way your skirt fits or whether your hair looks okay. You can't challenge power structures if gaining a few pounds leaves you feeling worthless. It's hard to stand up for what's right if your insides are crumbling under the weight of feeling ugly and invisible. When you're beauty sick, your batteries will always be running a bit low. You will never be at full power.

I can think of no one more suited than Colleen to share a story of how letting go of beauty sickness can free you to make your mark on the world. Today Colleen touches lives wherever she goes, leaving a trail of positive emotions in her wake. But she hasn't always been this way. For several years, every ounce of energy Colleen had was dedicated to racking up miles on the treadmill or calculating calories. During this period of time, she gave everything to her eating disorder. Trapped in the mirror, she had nothing left for the world around her.

Colleen is a twenty-four-year-old white woman. She lives in

Washington, DC, but is originally from North Carolina. Colleen bleeds Carolina blue, describing herself as a "Tar Heel: born, bred, graduated, dead." Her voice still carries a southern twang, particularly when she's excited about something, which is often. She burst into my office one early spring day after we'd spent nearly two months trying to sync our schedules to find a time to meet.

When she first contacted me, Colleen was a graduate student focusing on strategic communications. She said she wanted to talk with me about body image. But it ends up that Colleen didn't want to just talk about body image, she wanted to *do something* about body image. Colleen's particular concern was how fitness settings often encourage body shame in women. She wanted to know if I had specific evidence she could share with fitness instructors to prove to them that they shouldn't focus on appearance when teaching and training. I told her I didn't have that type of data, but discussed the sorts of studies that would be necessary to obtain it. Colleen's response? "Great. When do we get started?"

By the end of our first meeting, I had committed to conducting two different research studies with Colleen and she somehow convinced her adviser in the school of journalism to approve our work together as an independent study course. Colleen is someone who makes things happen, and she often makes them happen so quickly it can leave your head spinning.

Colleen is a whirlwind of energy and enthusiasm. She races through her words with a breathless intensity. She smacks her hands on the table for emphasis. She throws her head back in raucous laughter. We joked about how the transcriptionist for her interview might have a difficult time coming up with a way to account for the variety of her laughs.

The Colleen you see today is not so different from the Colleen you would have met when she was a child. She spent most of her early years blissfully free of the type of body worries that plague so many girls and women. She played sports every season, but wasn't super competitive. She ate what she wanted. She just didn't really think about how she was presenting herself or her body to other people. The only appearance-related thoughts she really remembers having as a child had to do with her eye color. Colleen's eyes are a piercing blue. I describe them as Carolina blue, which makes her beam.

"I remember as a kid getting very irritated because my sister and I had very blue eyes," Colleen explained. "Both of us did. And the only thing people would ever say to us is, 'Oh, you have such beautiful eyes!' It was irritating."

"Why did that irritate you?" I asked.

"I was just sick of hearing it. I was so sassy as a kid and such a tomboy. I would have loved to hear things like 'Oh, you're so funny' or 'You're so much fun' or 'You're so brave.'"

Colleen is articulating a message our culture is just starting to hear. When we limit ourselves to praising young girls for how they look, we're sending a message that the other pieces of them matter less. Even worse, we can inadvertently suggest that we don't even see the other components that make them who they are. Young Colleen must have wondered if people didn't notice how brave and funny she was. Didn't those things matter more than her eye color? If they did, why was eye color all she ever heard about?

Beyond comments about her eye color, Colleen doesn't remember paying much attention to how she looked as a child. I asked Colleen, "When you were young, if someone had asked you, 'Do you like your body?' what would you have said?"

"I would have never thought about it," Colleen shrugged. "Literally. It would seem like a weird question."

Like most girls, Colleen gained weight as she went through puberty. And just as it does for too many girls, those bodily changes brought her a new, troubling perspective on her body. As she closed in on the end of high school, Colleen was the heaviest she'd been, though she wasn't overweight by any medical definition. Colleen began to lose that easy childhood feeling of being at home in her body. The feeling started small and foreign, but grew to take up more and more space in her mind.

Around that time, Colleen went on an Outward Bound trip with other high school students. At one point, they all decided to go swimming in a lake and the girls in the group stripped down to their sports bras. But Colleen froze. She stared at the other girls as they launched themselves into the water.

"I realized that I felt insanely uncomfortable in my body compared to them," Colleen remembers.

"What made you uncomfortable?" I asked.

"It was my stomach," Colleen responded with certainty. "I didn't care about anything else. I wasn't cognizant of my thighs. I wasn't cognizant of my butt, my shoulders, my hands, my arms, anything. It was my stomach. I remember thinking, 'I can't take my shirt off because I don't want anyone to see the rolls.'"

Looking back, Colleen admits she didn't even really have rolls per se, just enough to grab. But that's not really the point. What's more important is that Colleen recalls this as the first moment she thought, "Other people will be looking at my body and I am worried about that." It's the first moment she consciously started to think of her body as something for others to evaluate.

As Colleen saw her body more and more through the eyes of others, she started to experience her body as an object instead of an instrument. Consistent with the research presented throughout this book, that feeling of being looked at interrupted her thoughts and her actions. It pulled her out of the present and left her locked in a mental mirror.

Colleen did go swimming with her friends that day, but she kept her shirt on. She has a souvenir from then, what she describes as "a really awkward-looking photo of me with my shirt on and everyone else with theirs off."

"If you could go back to young Colleen in that moment, what would you want to tell her?" I asked.

"I would totally have told her to take that shirt off!" Colleen's voice increased in volume so much that a colleague peeked into my office to see what was going on.

Colleen continued, uninterrupted. "I would have been, like, 'You are in the middle of a national forest having one of the most life-changing experiences. Just enjoy this moment and stop worrying about it. You have this incredible adventure, and what you're perseverating on is how your body looks to other people who don't even give a shit. Of all the qualities you could be focusing on this trip—on building confidence, going out of your comfort zone, developing your skills as a rock climber, team building—you're focused on your stomach. And it's just, it's sad.'"

The next big turning point for Colleen was when she went to college. She signed up for a course called "Intermediate Jogging." I had to laugh at this. When I was in college, we also had the option of taking those types of courses. I described for Colleen how I fumbled my way through "Beginning Tumbling" as a sophomore. Colleen's school, the University of North Carolina, mandated at least one exercise class, though students could pick the type of exercise.

One of the activities in Colleen's jogging class required students to weigh themselves at the beginning of the week, track all their exercise and food for the week, estimate calories in and calories out, then weigh again at the end of the week. Colleen describes herself as a perfectionist with a ridiculous work ethic. So when she approached this activity, she was determined to do a good job at it, just as she would any other class assignment. Despite the fact that Colleen was extremely fit at the time, she said the instructor of the class told her that her BMI was in the borderline range.

Colleen remembers thinking, "Oh my god, there's something wrong with me. I have to work harder." And so she did, using the app provided by the class for that initial tracking activity. "I'm going to exercise as much as I can and eat as little as I can to make the deficit go low, because the app rewards you when you have deficits," Colleen explained. It became a challenge to see how big she could make her deficit. She'd go multiple days in a row seeing nothing but deficit, deficit, deficit. Colleen continued this behavior for over a year. At the peak of her struggle, she was spending six hours per day in the gym.

Just as it did for Rebecca, the competitive swimmer you met in chapter 4, an ill-advised comment about BMI from a health instructor set Colleen down the road to systematically destroying her own health. Colleen kept tracking long after her assignment ended, in much the same way that Rebecca spent years being ruled by the mental spreadsheet her health teacher planted in one brief moment in one class.

When Colleen went home for the summer, her weight loss garnered praise and compliments. People would say, "Oh my god, what happened to you? You look really good." She remembers thinking, "Wow. I'm doing something really right." So she did more of it.

Colleen and I both agree that this is one of many reasons to tamp down appearance commentary directed at women. You never know what it means to a woman when you compliment her body. Praise that might seem simple on the surface can be loaded. When you praise her for losing weight, does that mean she was unacceptable before she lost weight? Does it mean that other women who are heavier than she is are not attractive? When you tell a woman you envy what you see as her ability to stay thin with little effort, do you have any idea whether that woman may be like Nique from chapter 12, exhausted from hearing others tell her she needs to gain weight? Or whether she might be struggling with an eating disorder? Or whether another woman who's listening to you is?

Once when I was in graduate school, I got a terrible case of the flu and dropped a good deal of weight in a short amount of time. When I returned to campus, a professor said, "You look good! Did you lose weight?" When I responded that I had lost weight because I'd been seriously ill, she just shrugged and said, "Well, however it happened, looks good!" I remember that moment as such a clear example that much of what we claim to be health-based concern about other women's weight is not at all. It's nothing more than an ill-disguised bit of buy-in to a culture that says our worth is determined by our body size and that less is always more, no matter how we get there.

The same summer that Colleen got those weight-loss compliments, she also learned that one of her archenemies from elementary and middle school, a girl she referred to as "the Regina George of Claxton Elementary," had gained weight in college. Colleen explained, "I absolutely hated this girl. Her mission in elementary and middle school and part of high school was to torture me. I'm still salty about her."

"I can tell!" I said.

"She gained a lot of weight in college," Colleen continued. "And so at the same time people were praising me for losing weight, they were making fun of her."

"And something felt good about that?" I asked gently.

Colleen looks down. "I am the most ashamed about this of anything that I have ever done or said in my life, but yeah. It felt awesome. And I remember feeling like I had won, and that I was better than her. And that's so embarrassing to me and that's so shameful, especially considering everything I stand for now."

It's hard for me to imagine Colleen glorying in someone else's psychological pain, because her focus on building others up is so thorough these days. But I believe everything she is saying. Her shame is visible.

During that period, Colleen had energy only for her disorder. She explains, "I would avoid social situations and spend my whole life in the gym. I'd wake up, I'd go to the gym, I'd go to class, I'd go to the gym, go to classes, come back, eat dinner, go back to the gym, go to bed." Colleen derived no pleasure from all that exercise. She followed strict rules, sticking mostly to cardio, and then calling any strength training she did "extra credit."

Colleen looks like she's horrified that she would ever have approached fitness in that way. I asked her if that was true.

"Yes!" she responded. "I am horrified! Horrified." Though her disorder will always be a piece of her, it's almost as though she is talking about a different person when she describes this time in her life. That makes sense in a way, because as Colleen described it, "when you have an eating disorder, you lose all sense of your own personality or your values."

Colleen became more and more isolated. Social events were difficult because she had created such a strict set of rules around what she could and could not eat. She'd make up excuses, telling people she was lactose intolerant or couldn't eat sugar. Though Colleen's case is extreme, it has clear parallels to the chronic dieting so many women slog through in the hopes of building a "dream body." Excessive food restrictions take up your mental energy and take you away from those you care about. It's harder to connect with others when your brain is too busy charting and planning every calorie.

Colleen summed it up. "I am fun and social and outgoing and loud, and all of that had been completely lost. I was spending all my time in a gym, and if I wasn't in the gym, I was thinking about exercising, I was thinking about food, and I couldn't hold a conversation because that's where my brain was. So it was almost futile to hang out with me, because I was so distracted."

"I can't even fathom that version of you." I said.

Colleen nodded and continued, "I remember a night I couldn't sleep. I couldn't sleep, I couldn't sleep, I couldn't sleep, and I realized I couldn't sleep because I was so hungry that my body wouldn't let me fall asleep. I got up, I ate a pear, and I fell asleep immediately. It's one of those things where I realized I had completely lost connection with my body and its senses."

"Like you didn't know what it needed or what it wanted?" I asked.

"Not at all," Colleen agreed. "No idea at all."

Colleen's talking about the concept of interoceptive awareness, first introduced in chapter 4. Interoceptive awareness makes you sensitive to what your body needs and how it feels moment to moment. Colleen's body had become nothing more than a pleasing or displeasing shape. She was treating her body as a performance object rather than a means of interacting with the world. The more it became an object, the less she was even aware of its actual needs. The comments she got reinforcing her weight loss didn't help. But even worse, when Colleen was wasting away, some still found a way to make her feel too heavy.

In the throes of her disorder, Colleen purchased a new dress that she described as "gorgeous, gorgeous, gorgeous."

A woman told Colleen she looked wonderful in the dress, but then said, "For how skinny you are, you have a pooch." Colleen says that comment destroyed her. The woman tried to backtrack, saying,

"Oh, it's because you eat so many vegetables! It's just gas!" But that made Colleen even more embarrassed.

Colleen was working at the front desk in a gym during this time period. One of her favorite trainers, who was moving away, stopped by the desk on her last day. She looked Colleen in the eyes and said, "Colleen, please take care of yourself."

That comment felt just as healing to Colleen as those body-shaming comments felt destructive. Colleen explained, "Everything that needed to be said was communicated in that very, very short moment, and in such a compassionate way. And it was in that moment that I realized that I had a problem." That woman's kindness, combined with Colleen's father's insistence that she start eating, and a well-timed and determined intervention by two of Colleen's close friends, finally resulted in Colleen getting help.

There was no magic, immediate cure. Colleen describes the work of leaving her eating disorder behind as "an excruciating climb." But she realized she wasn't alone during that climb. Other women were scaling that same mountain, and Colleen couldn't stop thinking about how she could help them.

Colleen never forget those words: "Please take care of yourself." She heard from loved ones who wanted to help her take care of herself, but often didn't know what to say or do. So Colleen decided that if she was going to make this excruciating climb, she would put a system in place to make it easier for those who might follow. She got together with some friends and started to design a program to train students how to help those struggling with eating disorders. She described the beginning of this movement as a little chaotic. No one really knew how to make the plan happen, but Colleen wouldn't give up. "We'd have guerrilla meetings in empty classrooms, talking about how to get the training going." Eventually, Colleen and her friends partnered with the Center of Excellence for Eating Disorders at UNC, and the program became a reality. This movement Colleen spearheaded is now called Embody. It's in the early stages of a nationwide rollout.

Colleen needed the activism of Embody to push her recovery forward. She explains, "Embody gave me ownership over my own situation. And not only my own situation, but the ability to help other people. I realized how important it was to take care of myself. Because I can't help people if I'm still sick. I wanted people to see me as this strong, recovered person."

"So you just had to be that person then?" I asked.

"I had to be that person." Colleen agreed.

Beauty sickness turns us away from the world and drains our compassion. It leaves us stuck in our heads, bound by our reflections. One of the best ways to break that cycle is to reach a hand out to others. There's no room for body-based rumination when your thoughts are targeted toward being there for other people. When you build compassion for others, some of that compassion will always stay within you, slowly replacing shame and self-doubt with hope and clarity.

For Colleen, part of being a strong, recovered person meant that she had to find a new way to relate to exercise. It needed to be fun instead of being a punishment. It needed to be healthy instead of a means to fuel her disorder. When I asked Colleen how she went from using exercise in a disordered way to being a fitness instructor who advocates for a body-positive approach to exercise, she laughed. "It's a funny story," she began.

A fellow student in one of Colleen's classes asked Colleen if she'd ever tried Zumba. She said, "It's Latin dance and aerobics!"

Colleen remembers thinking, "This is the stupidest shit I've ever heard in my life. This sounds like Jazzercise but stupider and there's no way in hell I'm doing it." But she thought going would be a good way to make friends, so she feigned enthusiasm and agreed to attend. "Sounds like a great time! Shake my ass in a fitness class? Absolutely!"

Colleen actually did have a great time. By her second Zumba class, she announced to her friend that she was going to become an instructor.

"I bet you were the most enthusiastic instructor ever!" I said.

Colleen told me she was equal parts enthusiastic and horrible in the beginning. But she found she loved teaching Zumba. She expanded her repertoire to include other types of classes and got a job at a local gym. But all the while she was preaching a healthy approach to others, her old attitudes and behaviors still lingered in the background.

When Colleen was at that gym, all of her coworkers told her she had to take a class by an instructor named Melanie. People talked about Melanie like she was a goddess and, Colleen said, "the whole place was haunted with her." Colleen finally decided to give it a try.

"So I go to this class and I'm sitting around and I'm looking around the room for, like, Jillian Michaels. I'm looking for this badass lady that's going to kick my ass."

"Like she's going to have a six-pack and killer biceps?" I ask.

"Oh yeah," Colleen continues. "Ripped as hell. But the room was full of pretty normal-looking people. There wasn't anybody in the room that stood out as particularly athletic looking. But as the class started, a heavier woman with frizzled hair and glasses, definitely no Jillian Michaels, gets up in front of the room, smiles, and says, 'Hi, I'm Melanie.' and I'm thinking, 'No way. There's no way in hell.' Never have I eaten my own words so quickly." Colleen slows her speech at the end, seeming to fit a few extra syllables into the word *never*.

Despite her initial misgivings, after one class, Colleen declared Melanie "the biggest badass on the face of this planet."

Melanie forced Colleen to confront her own fat phobia and the prejudice she often felt toward fat people. Colleen explained, "If somebody saw her, they would not understand the way that these people care about her, the way these people love her, the community she's created, how hard she works. There is nothing about this woman that is not incredible, and she is as kind and gracious as she is strong. And that's who I want to be. And I would rather be a Melanie than a Jillian Michaels any day. What I needed was a role model who showed me that fitness goes far beyond what you look like. Because everybody tells you the opposite."

Colleen pointed out that the advertisements you see for gyms and fitness classes are misleading and incomplete. "You can't see how the instructor makes people feel. On a poster, you can't see kindness. You can't see community building, you can't see motivation. All you can see is an amazing six-pack and a sports bra and booty shorts. And that somehow embodies the ideal of hard work and determination." Colleen makes a noise as if to say, "What bullshit."

Colleen began to channel Melanie in the fitness classes she was teaching. "I would discourage my students from overtraining, I would tell them to be sure that they were really well hydrated and fueling their bodies, and I did everything right, because I believed it for them. I never used negative body talk." Eventually, Colleen believed it for herself too. In healing others, she healed herself.

When I first met Colleen, what she really wanted from me was data to help persuade fitness instructors and trainers to take a less appearance-focused approach to their work. I didn't have that data, but with Colleen's help, we found a way to get it. We recruited over 200 women to take a sixteen-minute fitness class, filling out a few surveys immediately before and after the class. For this study,

Colleen taught the same class with the same exercises each time. However, we randomly assigned the women to one of two instructor scripts for the class.

In the appearance-focused class, Colleen used verbal cues focused on how the women looked or how they might like to change the appearance of their bodies. During an abdominal series, she'd say, "The movements we're doing are designed to blast fat and help you get those elusive six-pack abs. We'll be ready for bikinis in no time!" In the function-focused class, that comment would change to "These muscles are crucial to everything you do. They give you power! You're super strong and you have a lot of awesome things to do, so let's get to work!"

Colleen would be the first to tell you that it was not easy for her to teach an appearance-focused class. I'm grateful that she believed in the scientific method enough to do so for this study. As a minor rebellion, when she'd send me the audio recordings from each of the classes she taught for the study, she'd label the function-focused recording "nice" and the appearance-focused recording "asshole." I still laugh when I see those recordings in my files.

The good news is that Colleen's efforts paid off. We got solid data supporting her usual approach to fitness. The appearance-focused class led to increased self-objectification and lower body satisfaction in women. The function-focused class left women feeling happier and more body satisfied. When asked to describe the class in three words, those in the appearance-focused condition used words like *ashamed*, *sad*, *fat*, *weak*, and *self-conscious*. Those in Colleen's function-focused class used words like *motivated*, *strong*, *proud*, *accomplished*, and *enthusiastic*.

Colleen's using the evidence from this study and others to create a formal training program for fitness professionals that demonstrates how and why to teach nonobjectified classes that focus on appreciating what your body can do. She hopes to break down the idea that "your body is only as valuable as what it looks like." It's a perfect message for fitness classes, but that message can easily be expanded to something broader. We need to break down the notion that *women* are only as valuable as what they look like. When we no longer believe that, our lives will be healthier and freer. We'll be more able to tackle the world's challenges.

Colleen's perspective on fitness isn't just colored by her own years of overexercise or her experience teaching classes. There's another personal element to it as well. Colleen's mother, sister, and brother all have muscular dystrophy, a type that shows up later in life and results in the slow deterioration of muscles. It begins in the face and moves downward through the shoulders, arms, and legs. Colleen's mom will be in a wheelchair at some point. Her brother and sister likely will as well.

It's painful for Colleen to think back to the years when she failed to care for her body, when she never thought it was good enough. "I was abusing my body, and overusing it, and overexercising, and not appreciating it for the things that it could do. I was focusing so hard on appearance, when my brother and sister were fighting this battle of realizing that they were never, ever again going to be able to do things that most people can do." Colleen is currently training for a marathon with Team Momentum, a group that raises money for the Muscular Dystrophy Association.

"Doing the marathon with Team Momentum is really, really, really powerful because it makes you realize you should be grateful for all the things that your body can do. Because a lot of people don't have that." Colleen's voice quiets, then she repeats herself. "A lot of people don't have that."

Instead of worrying about how she looks, Colleen now appreciates what she calls "an incredible gift of physical vitality." She doesn't want to abuse that gift.

Appreciating what your body can do doesn't mean you need to be an athlete or spend every day at the gym. It certainly doesn't mean you have to be fully able bodied. Even if you feel your body is letting you down in some ways, it's still important to recognize the countless ways it's there for you. Colleen tells a beautiful story of how her mom still appreciates her own body, even as muscular dystrophy slowly makes its claim on it.

"One day my mom and I were walking up a mountain. And let it be known that this is not an easy mountain to walk up!" Colleen adds with pride. "I'm from Asheville, North Carolina, and shit is real hilly!"

She continues. "I remember walking it with my mom, and she said, 'I understand that I can't do everything I want to do. I can't play tennis. I can't do that. But, if my fitness is being able to walk up this mountain with my dog, I am happy with that.' And she said, 'Someday I might not be able to do that. Or maybe I'll be slower. But I will just embrace whatever I can do in that moment.'" Colleen smiles at the memory.

"How do you feel about your body now?" I asked Colleen.

She paused before answering. "Do I have bad days? Do I have lapses? Absolutely. There are times you slip and fall, but like anything else, you have two options. You move forward or you move backward. And I'm gonna continue to move forward no matter what."

FIGHTING BEAUTY SICKNESS IS ABOUT just that—moving forward. More important, it's about moving forward without the distracting presence of that mind-mirror. We deserve a chance to see what our lives might be like if we put that mirror down more often.

Beauty sickness means different things to different women. For some, it's the equivalent of a mild cold: annoying, but not serious. Other women's lives are so thoroughly disrupted by beauty sickness that, like Colleen, their focus on how they look can cause them to lose track of who they really are. Wherever you are on that spectrum, there is room for moving closer to the person you really want to be.

I was once asked by a student group to give a short speech focused on advice for college students. After struggling to come up with advice that would be useful or applicable to a wide range of students, I decided to alter my assigned topic. I've always been much more interested in asking questions than giving advice. So instead of delivering a traditional speech, I asked that group of students to consider a question: *What kind of person do you want to be?*

I've now asked that question of thousands of women. While their answers vary widely, not one has told me, "I want to be the kind of person others find pretty." Instead, they tell me they want to be the kind of people who bring joy and laughter to others, who heal the sick, who fearlessly explore new technologies, who nurture those who need nurturing, who create art that inspires, who write words that move, who fight for those who cannot fight for themselves. When I ask women another important question, *How do you want this world to be different when you leave it?*, they talk about fighting global warming or poverty or racism. They talk about leaving this place in better shape than they found it. If you ask these two questions of yourself, I doubt you will find answers that have much of anything to do with how you look.

Your responses to those two questions can be a first step toward loosening your grip on the mirror and the mirror's grip on you. Determine for you what matters most. What is it that you love? How do you want to spend your limited time and money? Where do you want to invest your limited emotional energy? If, after you answer those questions, you find you'd like to turn down the volume on beauty in order to turn up the volume on what matters more to you, consider some of these approaches for doing so.

Take a Beauty Inventory

First, let me be clear about something. I'm not suggesting that women abandon all beauty practices. That's neither realistic nor necessary. It's also not what most women would want. We're always going to care about how we look, and other people are too. That's not the problem. The problem is when caring about how we look moves us away from other important goals. You can care about how you look and still turn the beauty dial down a bit.

Unless we're deliberate about how we spend our beauty time and money, our beauty practices will control us instead of the other way around. Consider keeping a journal of how much time and money you're spending on beauty. Once you have your data, decide whether it makes sense to reallocate some of your resources and experiment with how it feels to do so.

The truth we often don't acknowledge is that no one cares about how we look quite as much as we do. You'll likely find that your world doesn't change for the worse as you imagine it might when you let some of that beauty time go. You might in fact find it changing for the better.

Be Gentle with Yourself

Women can't help but hear the casual cruelty with which our culture talks about their bodies. It's almost impossible not to internalize those voices to some extent. This leaves too many of us hearing our own voice saying we're not thin enough, not pretty enough, not good enough. I promise you that that type of inner monologue will not help you become healthier or happier. Practice putting a stop to that sort of self-talk.

The way to take better care of your body is by being kind to it and practicing gratitude for all the things it does for you. Just as we take good care of the people we love, we can learn to take good care of our bodies through practicing compassion instead of denigration. Don't listen to those in this world who claim you have to hate how your body looks in order to motivate healthy behaviors. That's nonsense. Those types of claims aren't supported by any type of scientific data. They're supported only by people making excuses for their unnecessary cruelty or self-righteousness. If you want to help a woman who is trying to improve the health of her body, the last thing you need to do is say anything at all about how she looks.

Move Toward Thinking About Your Body As Something That Does Instead of Something That Appears

A few years ago, I interviewed professional women's roller derby skaters for a research project. I still remember what one of those skaters said when I asked her how she thought about her body. She told me, "My body is a vessel, to do what I love." It's hard to remember that sometimes, when so much of what we see in our culture tells us that our bodies exist for others to look at and evaluate.

One of the clearest contexts in which to apply the "vessel" lesson is the context of working out. If you exercise, focus on exercising for pleasure, stress reduction, and improved health. Focusing on exercising to change how you look makes you less likely to stick with it. Avoid gyms, classes, or fellow exercisers that encourage you to work out for a new body shape. Instead, try to find instructors like Colleen and Melanie, who leave you feeling powerful instead of insecure. Exercise so that you'll be more able to do the things you want to do, not so that you can look the way people want you to look.

Another way to respect the fact that your body doesn't exist just for other people to look at is by dressing in a way that doesn't distract you from what you're doing. If you can't sit comfortably in an outfit, that's a guarantee that it will steal valuable brain space from you when you wear it.

Purchase clothes that make you feel like your best self, but ask yourself whether you can move comfortably in those clothes before you take them to the register. Make a commitment not to purchase clothing that requires constant monitoring or sucking in or uncomfortable poses to stay in place.

I think a lot about those girls I saw a few years ago in their homecoming dresses. I completely understand that for most girls, part of the fun of going to a school dance is dressing up and feeling attractive. I remember my own excitement at that age when I picked out a homecoming dress. Of course I wanted to look good in that dress. I don't think it's a big deal that those teenage girls wanted to feel attractive. But remember, these were the girls wearing dresses that were so short, they couldn't even sit comfortably at dinner. A few more inches of fabric could have bought back a good deal of brain space and still left those girls room to feel pretty for a special occasion. I'm fairly certain there's a happy medium within reach, a place where we can feel attractive but not be distracted by our own clothing. We have enough hurdles to jump over in life. We don't need to create more by selecting apparel that leaves us less able to be fully present in our lives.

Mind Your Media

If a destructive image or headline on a magazine catches your eye, practice immediately turning away (or turning it around!). If you feel yourself getting sucked into gossipy, beauty-focused television programs, articles, or websites, turn them off or click away. If you're posting pictures of yourself on social media, ask yourself why you're posting each picture and what you're trying to communicate with it. Make sure you're comfortable with the answers to those questions before you post. Social media likes are a poor balm for lack of confidence. A sexy picture is no cure for low self-esteem.

Watch Your Words

Maybe you don't want to change your beauty habits at all. Maybe you *like* spending time and money on your appearance. If you're happy with where you are, that's great. But I will still ask you to consider doing one particular thing to fight beauty sickness. Give other girls and women the freedom to feel they are more than what they look like by avoiding body talk and limiting appearance-related conversations.

Many of you may have heard of the Bechdel test. It first appeared in a comic strip created by Alison Bechdel in 1985. The Bechdel test poses three questions that help to determine whether a given piece of media (usually a film) includes women and represents their concerns beyond heterosexual romance. First, you ask whether the movie has at least two women in it. Then you determine whether those women actually speak to each other in the film. Finally, you ask whether what they speak to each other about is something other than a man.

I'd like to propose an extension of the Bechdel test that can be applied to conversations between women. "If two or more women are talking to each other, are they talking about something other than how they look?" I'm in no way trying to shame women who want to talk with each other about fashion or other appearance concerns. I'm certain there's a time and place for those conversations. But when all we hear other women talking about (whether in the media or in real life) is appearance, that sends the message that there's not much more to women than how they look. It suggests we don't have other things to talk about.

Appearance-driven conversations force every woman in hearing distance to think about her own appearance. Help the women you spend time with escape the internal mirror by encouraging conversations about other topics. Women talking about appearance with other women is a strong cultural norm. It's a quick and easy way to bond with a woman—just compliment her outfit or her haircut. It can be a hard habit to break, but it's worth giving it a try. Think back to your "what matters more" list. Remember that every woman has that list. Try to ask questions or pay compliments that address that list instead of feeding an already rampant focus on how women look.

Vote with Your Wallet

Marketers have long recognized that women drive the majority of purchase decisions—both via making purchases for themselves and through their influence on purchases made for others. This means that women are in a strong position to make change by voting with their wallets. Avoid supporting companies or brands that use destructive images or messages about women and girls to sell their products. More important, reward those companies and brands that send positive, healthy messages about women.

Don't hesitate to use your social media clout to call out advertisements that shame or belittle women. The last few years have been replete with examples of women making real change through the power of social media. In 2013, Harrods department store was the focus of a fast-moving tweetstorm in response to two books it displayed in its toy department. A book featuring a girl lounging on her bed was titled, *How to Be Gorgeous*. Its boy-targeted counterpart featured a triumphant boy behind a podium and was titled, *How to Be Clever*. In response to activism on social media, Harrods removed the books from its shelves.

GoDaddy, a leading website domain provider, provides another example of the positive results of this type of activism. After years of relying on sexist ads featuring scantily clad women, GoDaddy responded to thousands of #notbuyingit tweets along with pressure from women business owners. They abandoned their objectifying ad campaign and their CEO now speaks openly about fighting sexism.

When it comes to buying gifts for young girls, you have the potential to double your impact. In addition to strategically spending your dollars at companies that send positive messages about girls and women, you can also choose to give girls gifts that don't focus on appearance. Many little girls spend their birthdays or holidays being showered with presents that promote beauty sickness in one form or another. Consider choosing a gift that encourages girls to be brave or curious instead of just pretty. Girls already get enough messages about the value of their looks. You can be a different sort of gift giver.

Moving Forward

I chose to end *Beauty Sick* with Colleen's story, because it so clearly demonstrates the good that comes when a woman wins the battle to reclaim energy and time previously held captive by a relentless obsession with body shape. But Colleen also made an important point when we talked about how small steps can add up to something bigger.

"Do you think of yourself as someone who wants to change the world?" I asked Colleen, though I suspected I already knew the answer.

"Absolutely," Colleen confirmed, drawing out each vowel in the word. "If all I do is tell people, 'Your dreams and desires are more important than what society's expectations of you are,' then that in itself is a radical act against society's expectations. And it's a very simple and a very easy thing to do, so if I can do it, I will do it."

Then Colleen made an important point: A collection of sometimes seemingly minor incidents can create a larger, more damaging whole. Colleen used the example of media images to illustrate this idea. One magazine image doesn't wreck a young girl's body image. One overheard fat talk conversation doesn't cause an eating disorder. One social media post doesn't shut down your life goals. But, as Colleen put it, "It's the accumulation of a whole host of things that are contributing to this issue."

Just as a thousand tiny cuts from a beauty-sick culture can break a girl or woman down, a thousand small steps toward something better can build girls and women up. We can make meaningful cultural change by taking steps in our own lives to lessen the focus on women's appearance and by encouraging others to do the same. We can make even greater change when we work to hold organizations accountable for objectifying behaviors or ad campaigns. We can both offer more to the girls and women around us and demand more from our culture at large.

If you can imagine a world where girls and women are less objectified and do less self-objectification, you'll see a world where everything has changed. We would do different things. We would feel more ourselves and less defined by how much others enjoy looking at us. Our money and time would be spent differently. Our bodies would be healthier. Depression and anxiety might be less common or less severe.

It's time to focus on looking outward rather than being looked at. There's a lot to see out there in the world. There's a lot of work to be done.

Acknowledgments

WRITING THIS BOOK would have been impossible without the generosity and encouragement of so many students, colleagues, friends, and family members. I'm particularly grateful to those who read and provided feedback on early drafts of chapters, including Liz Morey Campbell, David Condon, Colleen Daly, Alice Eagly, Amberly Panepinto, and Jennifer Piemonte. Thanks also to my colleagues Bill Revelle (for the statistics and the stories) and Dan McAdams (for never failing to respond to my requests for advice).

Thank you to Northwestern University for the time, space, and support necessary to work on this book.

To my "shadow faculty" friends, who know my neuroses and like me anyway, thank you for listening. To the Nerd, who edited every single chapter, often multiple times, you are owed endless chocolate bars and undying gratitude.

I am forever indebted to the many talented members, past and present, of the Body and Media Lab at Northwestern University. BAMs, you inspire me every day by caring about things that matter and allowing that care to guide your actions. Thank you also to all of my past and present students, both at Loyola University and Northwestern University. You've taught me more than you realize.

You keep me on my toes, and I see you leading the way to a brighter future.

My mom is owed a special note of thanks for agreeing to let me tell the world that she applied makeup while driving me to school every morning. I was lucky never to have crashed, but am even luckier to have a mom who laughs easily and often.

To all of the baristas at the Starbucks on Sherman and Clark, thank you for knowing my name, remembering how I like my tea, and never seeming to mind the many hours I spent monopolizing my favorite spot, pounding away on my laptop. The vast majority of this book was written with your smiles and kind words as background music.

Thank you to the team at HarperCollins, who took a chance on a first-time author and led me through the process, especially Lisa Sharkey, Alieza Schvimer, and Amanda Pelletier.

It's no hyperbole to say that this book wouldn't have happened without Marcy Posner of Folio Literary Agency, who believed I should write a book well before I did. Thanks for being my pit bull, Marcy.

Most important, thank you to the girls and women who were brave enough to share their stories with me. Your words matter.

Notes

Chapter 1: Will I Be Pretty?

1. Damiano SR, Paxton SJ, Wertheim EH, McLean SA, Gregg KJ. Dietary restraint of 5-year-old girls: Associations with internalization of the thin ideal and maternal, media, and peer influences. *International Journal of Eating Disorders.* 2015;48(8):1166–1169.
2. Dohnt H, Tiggemann M. The contribution of peer and media influences to the development of body satisfaction and self-esteem in young girls: A prospective study. *Developmental Psychology.* 2006;42(5):929.
3. Shapiro S, Newcomb M, Burns Loeb T. Fear of fat, disregulated-restrained eating, and body-esteem: Prevalence and gender differences among eight- to ten-year-old children. *Journal of Clinical Child Psychology.* 1997;26(4):358–365.
4. Bearman SK, Martinez E, Stice E, Presnell K. The skinny on body dis-satisfaction: A longitudinal study of adolescent girls and boys. *Journal of Youth and Adolescence.* Apr 2006;35(2):217–229.

Chapter 2: Just Like a Woman

1. Rodin J, Silberstein L, Striegel-Moore R. Women and weight: A normative discontent. Paper presented at the Nebraska Symposium on Motivation. *Nebraska Symposium on Motivation.* 1984;32:267–307.
2. Feingold A, Mazzella R. Gender differences in body image are increasing. *Psychological Science.* 1998;9(3):190–195.
3. Inchley J et al., eds. Growing up unequal: Gender and socioeconomic differences in young people's health and well-being. *Health Behavior in School-Aged Children Study: International Report from the 2013/2014 Survey.* Vol. 7. Copenhagen: WHO Regional Office for Europe, 2016.

4. Bearman SK, Presnell K, Martinez E, Stice E. The skinny on body dissatisfaction: A longitudinal study of adolescent girls and boys. *Journal of Youth and Adolescence.* 2006;35(2):217–229.

5. Frederick DA, Peplau LA, Lever J. The swimsuit issue: Correlates of body image in a sample of 52,677 heterosexual adults. *Body Image.* 2006;3(4):413–419.

6. Gabriel MT, Critelli JW, Ee JS. Narcissistic illusions in self-evaluations of intelligence and attractiveness. *Journal of Personality.* 1994;62(1):143–155.

7. Halliwell E, Dittmar H. A qualitative investigation of women's and men's body image concerns and their attitudes toward aging. *Sex Roles.* 2003;49(11–12):675–684.

8. Fredrickson BL, Roberts T-A, Noll SM, Quinn DM, Twenge JM. That swimsuit becomes you: Sex differences in self-objectification, restrained eating, and math performance. *Journal of Personality and Social Psychology.* 1998;75(1):269.

9. Furnham A, Badmin N, Sneade I. Body image dissatisfaction: Gender differences in eating attitudes, self-esteem, and reasons for exercise. *Journal of Psychology.* 2002;136(6):581–596.

Chapter 3: I, Object

1. Mulvey L. Visual pleasure and narrative cinema. *Screen.* 1975;16:6–8.

2. Fredrickson BL, Roberts TA. Objectification theory:Toward understanding women's lived experiences and mental health risks. *Psychology of Women Quarterly.* June 1997;21(2):173–206.

3. Kozee HB, Tylka TL, Augustus-Horvath CL, Denchik A. Development and psychometric evaluation of the Interpersonal Sexual Objectification Scale. *Psychology of Women Quarterly.* June 2007;31(2):176–189.

4. LaForce M. Unpopular opinion: Gimme more blurred lines. *Thought Catalog, August 22, 2013.*

Chapter 4: Your Mind on Your Body and Your Body on Your Mind

1. McKinley NM, Hyde JS. The Objectified Body Consciousness Scale: Development and validation. *Psychology of Women Quarterly.* 1996;20(2):181–215.

2. Fredrickson BL, Roberts TA, Noll SM, Quinn DM, Twenge JM. That swimsuit becomes you: Sex differences in self-objectification, restrained

eating, and math performance. *Journal of Personality and Social Psychology.* July 1998;75(1):269–284.

3. Quinn DM, Kallen RW, Cathey C. Body on my mind: The lingering effect of state self-objectification. *Sex Roles.* December 2006;55(11–12):869–874.

4. Quinn DM, Kallen RW, Twenge JM, Fredrickson BL. The disruptive effect of self-objectification on performance. *Psychology of Women Quarterly.* March 2006;30(1):59–64.

5. Gapinski KD, Brownell KD, LaFrance M. Body objectification and "fat talk": Effects on emotion, motivation, and cognitive performance. *Sex Roles.* May 2003;48(9–10):377–388.

6. Martin KA. Becoming a gendered body: Practices of preschools. *American Sociological Review.* August 1998;63(4):494–511.

7. Murnen SK, Greenfield C, Younger A, Boyd H. Boys act and girls appear: A content analysis of gender stereotypes associated with characters in children's popular culture. *Sex Roles.* 2016;74(1–2):78–91.

8. Fredrickson BL, Harrison K. Throwing like a girl: Self-objectification predicts adolescent girls' motor performance. *Journal of Sport & Social Issues.* 2005;29(1):79–101.

9. Myers TA, Crowther JH. Is self-objectification related to interoceptive awareness? An examination of potential mediating pathways to disordered eating attitudes. *Psychology of Women Quarterly.* 2008;32(2):172–180.

Chapter 5: It's a Shame

1. Grabe S, Hyde JS, Lindberg SM. Body objectification and depression in adolescents: The role of gender, shame, and rumination. *Psychology of Women Quarterly.* June 2007;31(2):164–175.

2. Monro F, Huon G. Media-portrayed idealized images, body shame, and appearance anxiety. *International Journal of Eating Disorders.* 2005;38(1):85–90.

3. Cramer P, Steinwert T. Thin is good, fat is bad: How early does it begin? *Journal of Applied Developmental Psychology.* 1998;19(3):429–451.

4. Davison KK, Birch LL. Predictors of fat stereotypes among 9-year-old girls and their parents. *Obesity Research.* 2004;12(1):86–94.

5. Roehling MV, Roehling PV, Pichler S. The relationship between body weight and perceived weight-related employment discrimination: The

role of sex and race. *Journal of Vocational Behavior.* 2007;71(2):300–318.

6. Jasper CR, Klassen ML. Stereotypical beliefs about appearance: Implications for retailing and consumer issues. *Percept Motor Skill.* 1990;71(2):519–528.

7. Pingitore R, Dugoni BL, Tindale RS, Spring B. Bias against overweight job applicants in a simulated employment interview. *Journal of Applied Psychology.* 1994;79(6):909.

8. Miller BJ, Lundgren JD. An experimental study of the role of weight bias in candidate evaluation. *Obesity.* 2010;18(4):712–718.

9. Sheets V, Ajmere K. Are romantic partners a source of college students' weight concern? *Eating Behaviors.* 2005;6(1):1–9.

10. Vartanian LR, Shaprow JG. Effects of weight stigma on exercise motivation and behavior: a preliminary investigation among college-aged females. *Journal of Health Psychology.* 2008;13(1):131–138.

11. Sutin A, Robinson E, Daly M, Terracciano A. Weight discrimination and unhealthy eating-related behaviors. *Appetite.* July 2016;102:83–89.

12. Neumark-Sztainer D, Falkner N, Story M, Perry C, Hannan PJ, Mulert S. Weight-teasing among adolescents: correlations with weight status and disordered eating behaviors. *International Journal of Obesity and Related Metabolic Disorders.* 2002;26(1):123–131. Haines J, Neumark-Sztainer D, Eisenberg ME, Hannan PJ. Weight teasing and disordered eating behaviors in adolescents: Longitudinal findings from Project EAT (Eating Among Teens). *Pediatrics.* 2006;117(2):e209–e215. Fairburn CG, Welch SL, Doll HA, Davies BA, O'Connor ME. Risk factors for bulimia nervosa: A community-based case-control study. *Archives of General Psychiatry.* 1997;54(6):509–517. Storch EA, Milsom VA, DeBraganza N, Lewin AB, Geffken GR, Silverstein JH. Peer victimization, psychosocial adjustment, and physical activity in overweight and at-risk-for-overweight youth. *Journal of Pediatric Psychology.* 2007;32(1):80–89.

13. Sutin A, Robinson E, Daly M, Terracciano A. Weight discrimination and unhealthy eating-related behaviors. *Appetite.* July 2016;102:83–89.

14. Arcelus J, Mitchell AJ, Wales J, Nielsen S. Mortality rates in patients with anorexia nervosa and other eating disorders. A meta-analysis of 36 studies. *Archives of General Psychiatry.* July 2011;68(7):724–731.

15. Berg KC, Frazier P, Sherr L. Change in eating disorder attitudes and

behavior in college women: Prevalence and predictors. *Eating Behaviors.* 2009;10(3):137–142.

16. Laberg JC, Wilson GT, Eldredge K, Nordby H. Effects of mood on heart rate reactivity in bulimia nervosa. *International Journal of Eating Disorders.* 1991;10(2):169–178.

17. Green MW, Rogers PJ, Elliman NA, Gatenby SJ. Impairment of cognitive performance associated with dieting and high levels of dietary restraint. *Physiology & Behavior.* March 1994;55(3):447–452.

18. Tiggemann M. Dietary restraint as a predictor of reported weight loss and affect. *Psychological Reports.* December 1994;75(3 Pt 2):1679–1682.

19. Mann T, Tomiyama AJ, Westling E, Lew A-M, Samuels B, Chatman J. Medicare's search for effective obesity treatments: Diets are not the answer. *American Psychologist.* 2007;62(3):220.

20. Wildman RP, Muntner P, Reynolds K, et al. The obese without cardiometabolic risk factor clustering and the normal weight with cardiometabolic risk factor clustering: Prevalence and correlates of 2 phenotypes among the US population (NHANES 1999–2004). *Archives of Internal Medicine.* 2008;168(15):1617–1624.

21. Bazzini DG, Pepper A, Swofford R, Cochran K. How healthy are health magazines? A comparative content analysis of cover captions and images of women's and men's health magazine. *Sex Roles.* 2015;72(5–6):198–210.

22. Hankin BL, Abramson LY, Moffitt TE, Silva PA, McGee R, Angell KE. Development of depression from preadolescence to young adulthood: emerging gender differences in a 10-year longitudinal study. *Journal of abnormal psychology.* 1998;107(1):128.

23. Grabe S, Hyde JS, Lindberg SM. Body objectification and depression in adolescents: The role of gender, shame, and rumination. *Psychology of Women Quarterly.* June 2007;31(2):164–175.

24. Brausch AM, Muehlenkamp, JJ. Body image and suicidal ideation in adolescents. *Body Image.* 2007;4(2):207–212.

25. Miner-Rubino K, Twenge JM, Fredrickson BL. Trait self-objectification in women: Affective and personality correlates. *Journal of Research in Personality.* 2002;36(2):147–172.

26. Muehlenkamp JJ, Saris–Baglama RN. Self-objectification and its psychological outcomes for college women. *Psychology of Women Quarterly.* 2002;26(4):371–379.

27. Impett EA, Henson JM, Breines JG, Schooler D, Tolman DL. Embodiment feels better: Girls' body objectification and well-being across adolescence. *Psychology of Women Quarterly.* March 2011;35(1):46–58.

Chapter 6: Your Money and Your Time

1. YWCA. Beauty at any cost. 2008; http://www.ywca.org/atf/ cf/%7B711d5519–9e3c-4362-b753-ad138b5d352c}/BEAUTY-AT-ANY-COST.PDF.
2. 100 Million Dieters, $20 Billion. ABC News. 2012; http://abcnews.go.com/ Health/100-million-dieters-20-billion-weight-loss-industry/ story?id=16297197.
3. Mattingly MJ, Blanchi SM. Gender differences in the quantity and quality of free time: The US experience. *Social Forces.* 2003;81(3):999–1030.

Chapter 7: Malignant Mainstream Media

1. Banksy. *Cut It Out.* Weapons of Mass Disruption, 2004.
2. Fouts G, Burggraf K. Television situation comedies: Female body images and verbal reinforcements. *Sex Roles.* 1999;40(5–6):473–481.
3. Greenberg BS, Eastin M, Hofschire L, Lachlan K, Brownell KD. Portrayals of overweight and obese individuals on commercial television. *American Journal of Public Health.* 2003;93(8):1342–1348.
4. Stice E, Shaw HE. Adverse effects of the media portrayed thin-ideal on women and linkages to bulimic symptomatology. *Journal of Social and Clinical Psychology.* 1994;13(3):288–308.
5. Owen PR, Laurel-Seller E. Weight and shape ideals: Thin is dangerously in. *Journal of Applied Social Psychology.* 2000;30(5):979–990.
6. Ballentine LW, Ogle JP. The making and unmaking of body problems in *Seventeen* magazine, 1992–2003. *Family and Consumer Sciences Research Journal.* 2005;33(4):281–307.
7. Wasylkiw L, Emms A, Meuse R, Poirier K. Are all models created equal? A content analysis of women in advertisements of fitness versus fashion magazines. *Body Image.* 2009;6(2):137–140.
8. Groesz LM, Levine MP, Murnen SK. The effect of experimental presentation of thin media images on body satisfaction: A meta-analytic review. *International Journal of Eating Disorders.* 2002;31(1):1–16.

9. Harrison K. Television viewers' ideal body proportions: The case of the curvaceously thin woman. *Sex Roles.* 2003;48(5–6):255–264.

10. Sperry S, Thompson JK, Sarwer DB, Cash TF. Cosmetic surgery reality TV viewership: Relations with cosmetic surgery attitudes, body image, and disordered eating. *Annals of Plastic Surgery.* 2009;62(1):7–11.

11. Becker AE. Television, disordered eating, and young women in Fiji: Negotiating body image and identity during rapid social change. *Culture, Medicine and Psychiatry.* 2004;28(4):533–559. Becker AE, Hamburg P. Culture, the media, and eating disorders. *Harvard Review of Psychiatry.* 1996;4(3):163–167.

12. Goffman E. *Gender Advertisements.* New York: Harper and Row, 1979.

13. Copeland GA. Face-ism and primetime television. *Journal of Broadcasting & Electronic Media.* 1989;33(2):209–214. Archer D, Iritani B, Kimes DD, Barrios M. Face-ism: Five studies of sex differences in facial prominence. *Journal of Personality and Social Psychology.* 1983;45(4):725.

14. Smith LR, Cooley SC. International faces: An analysis of self-inflicted face-ism in online profile pictures. *Journal of Intercultural Communication Research.* 2012;41(3):279–296.

15. Archer et al. Face-ism. 45(4):725.

16. Bernard P, Gervais SJ, Allen J, Campomizzi S, Klein O. Integrating sexual objectification with object versus person recognition the sexualized-body-inversion hypothesis. *Psychological Science.* 2012;23(5): 469–471.

17. Milburn MA, Mather R, Conrad SD. The effects of viewing R-rated movie scenes that objectify women on perceptions of date rape. *Sex Roles.* 2000;43(9–10):645–664.

18. Yao MZ, Mahood C, Linz D. Sexual priming, gender stereotyping, and likelihood to sexually harass: Examining the cognitive effects of playing a sexually-explicit video game. *Sex Roles.* 2010;62(1–2):77–88.

19. Heflick NA, Goldenberg JL. Objectifying Sarah Palin: Evidence that objectification causes women to be perceived as less competent and less fully human. *Journal of Experimental Social Psychology.* 2009;45(3):598–601.

20. Cikara M, Eberhardt JL, Fiske ST. From agents to objects: Sexist attitudes and neural responses to sexualized targets. *Journal of Cognitive Neuroscience.* 2011;23(3):540–551.

21. Greenberg BS, Eastin M, Hofschire L, Lachlan K, Brownell KD. Portrayals

of overweight and obese individuals on commercial television. *American Journal of Public Health.* 2003;93(8):1342–1348. Fouts G, Burggraf K. Television situation comedies: Female weight, male negative comments, and audience reactions. *Sex Roles.* 2000;42(9–10):925–932.

Chapter 8: (Anti)social Media and Online Obsessions

1. Slater A, Tiggemann M, Hawkins K, Werchon D. Just one click: A content analysis of advertisements on teen web sites. *Journal of Adolescent Health.* 2012;50(4):339–345.

2. Boepple L, Thompson JK. A content analytic comparison of fitspiration and thinspiration websites. *International Journal of Eating Disorders.* 2016;49(1):98–101.

3. Custers K, Van den Bulck J. Viewership of pro-anorexia websites in seventh, ninth and eleventh graders. *European Eating Disorders Review.* 2009;17(3):214–219.

4. Chua THH, Chang L. Follow me and like my beautiful selfies: Singapore teenage girls' engagement in self-presentation and peer comparison on social media. *Computers in Human Behavior.* 2016;55:190–197.

5. Kapidzic S, Herring SC. Gender, communication, and self-presentation in teen chatrooms revisited: Have patterns changed? *Journal of Computer-Mediated Communication.* 2011;17(1):39–59.

6. Berne S, Frisén A, Kling J. Appearance-related cyberbullying: A qualitative investigation of characteristics, content, reasons, and effects. *Body Image.* 2014;11(4):527–533.

7. Lydecker JA, Cotter EW, Palmberg AA, et al. Does this Tweet make me look fat? A content analysis of weight stigma on Twitter. *Eating and Weight Disorders—Studies on Anorexia, Bulimia, and Obesity.* June 2016;21(2):229–235.

8. Buckels EE, Trapnell PD, Paulhus DL. Trolls just want to have fun. *Personality and Individual Differences.* 2014;67:97–102.

9. Fardouly J, Diedrichs PC, Vartanian LR, Halliwell E. Social comparisons on social media: The impact of Facebook on young women's body image concerns and mood. *Body Image.* 2015;13:38–45.

10. Manago AM, Ward LM, Lemm KM, Reed L, Seabrook R. Facebook involvement, objectified body consciousness, body shame, and sexual assertiveness in college women and men. *Sex Roles.* 2015;72(1–2):1–14.

Chapter 9: Media Literacy Is Not Enough

1. Selimbegović L, Chatard A. Single exposure to disclaimers on airbrushed thin ideal images increases negative thought accessibility. *Body Image.* 2015;12:1–5.

2. Paraskeva N, Lewis-Smith H, Diedrichs PC. Consumer opinion on social policy approaches to promoting positive body image: Airbrushed media images and disclaimer labels. *Journal of Health Psychology.* 2015:1359105315597052.

3. Nathanson AI, Botta RA. Shaping the effects of television on adolescents' body image disturbance: The role of parental mediation. *Communication Research.* 2003;30(3):304–331.

4. Botta RA. Television images and adolescent girls' body image disturbance. *Journal of Communication.* 1999;49(2):22–41.

5. Murnen SK, Smolak L. Are feminist women protected from body image problems? A meta-analytic review of relevant research. *Sex Roles.* 2009;60(3–4):186–197.

Chapter 10: The Problem with "Real Beauty"

1. Tiggemann M, Boundy M. Effect of environment and appearance compliment on college women's self-objectification, mood, body shame, and cognitive performance. *Psychology of Women Quarterly.* Dec 2008;32(4):399–405.

2. Calogero RM. A test of objectification theory: The effect of the male gaze on appearance concerns in college women. *Psychology of Women Quarterly.* Mar 2004;28(1):16–21.

Chapter 11: Turning Down the Volume

1. Langlois JH, Kalakanis L, Rubenstein AJ, Larson A, Hallam M, Smoot M. Maxims or myths of beauty? A meta-analytic and theoretical review. *Psychological Bulletin.* 2000;126(3):390.

2. Langlois JH, Ritter JM, Roggman LA, Vaughn LS. Facial diversity and infant preferences for attractive faces. *Developmental Psychology.* 1991;27(1):79.

3. Rhodes G, Yoshikawa S, Palermo R, et al. Perceived health contributes to the attractiveness of facial symmetry, averageness, and sexual

dimorphism. *Perception.* 2007;36(8):1244–1252. Nedelec JL, Beaver KM. Physical attractiveness as a phenotypic marker of health: An assessment using a nationally representative sample of American adults. *Evolution and Human Behavior.* 2014;35(6):456–463.

4. Cash TF, Melnyk SE, Hrabosky JI. The assessment of body image investment: An extensive revision of the Appearance Schemas Inventory. *International Journal of Eating Disorders.* 2004;35(3):305–316.

5. Diener E, Seligman ME. Very happy people. *Psychological Science.* 2002;13(1):81–84.

6. Stice E, Rohde P, Shaw H. *The Body Project: A Dissonance-Based Eating Disorder Prevention Intervention.* New York: Oxford University Press, 2012.

7. Stice E, Yokum S, Waters A. Dissonance-based eating disorder prevention program reduces reward region response to thin models: How actions shape valuation. *Plos One.* 2015;10(12):e0144530.

Chapter 12: Stop the Body Talk

1. Becker CB, Diedrichs PC, Jankowski G, Werchan C. I'm not just fat, I'm old: Has the study of body image overlooked "old talk"? *Journal of Eating Disorders.* 2013;1(1):1.

2. Neumark-Sztainer D, Falkner N, Story M, Perry C, Hannan PJ. Weight-teasing among adolescents: Correlations with weight status and disordered eating behaviors. *International Journal of Obesity and Related Metabolic Disorders.* 2002;26(1).

Chapter 13: Function over Form

1. Noll SM, Fredrickson BL. A mediational model linking self-objectification, body shame, and disordered eating. *Psychology of Women Quarterly.* 1998;22(4):623–636.

2. Cash TF, Novy PL, Grant JR. Why do women exercise? Factor analysis and further validation of the Reasons for Exercise Inventory. *Percept Motor Skill.* 1994. Tylka TL, Homan KJ. Exercise motives and positive body image in physically active college women and men: Exploring an expanded acceptance model of intuitive eating. *Body Image.* 2015;15:90–97.

3. Segar M, Spruijt-Metz D, Nolen-Hoeksema S. Go figure? Body-shape mo-

tives are associated with decreased physical activity participation among midlife women. *Sex Roles.* 2006;54(3–4):175–187.

4. Silberstein LR, Striegel-Moore RH, Timko C, Rodin J. Behavioral and psychological implications of body dissatisfaction: Do men and women differ? *Sex Roles.* 1988;19(3–4):219–232.

Chapter 14: Learning to Love Your Body and Teaching Others to Do the Same

1. Wood-Barcalow NL, Tylka TL, Augustus-Horvath CL. "But I like my body": Positive body image characteristics and a holistic model for young-adult women. *Body Image.* 2010;7(2):106–116.

2. Slater A, Tiggemann M. The influence of maternal self-objectification, materialism and parenting style on potentially sexualized 'grown up' behaviours and appearance concerns in 5–8 year old girls. *Eating Behaviors.* 2016;22:113–118.

3. Wansink B, Latimer LA, Pope L. "Don't eat so much": How parent comments relate to female weight satisfaction. *Eating and Weight Disorders—Studies on Anorexia, Bulimia, and Obesity.* 2016:1–7.

4. Coffman DL, Balantekin KN, Savage JS. Using propensity score methods to assess causal effects of mothers' dieting behavior on daughters' early dieting behavior. *Childhood Obesity.* 2016.

5. Pietiläinen K, Saarni S, Kaprio J, Rissanen A. Does dieting make you fat?; A twin study. *International Journal of Obesity.* 2012;36(3):456–464.

6. Perez M, Kroon Van Diest AM, Smith H, Sladek MR. Body dissatisfaction and its correlates in 5- to 7-year-old girls: A social learning experiment. *Journal of Clinical Child & Adolescent Psychology.* 2016:1–13.

7. Avalos L, Tylka TL, Wood-Barcalow N. The Body Appreciation Scale: Development and psychometric evaluation. *Body Image.* 2005;2(3):285–297.

8. Andrew R, Tiggemann M, Clark L. Positive body image and young women's health: Implications for sun protection, cancer screening, weight loss and alcohol consumption behaviours. *Journal of Health Psychology.* 2016;21(1):28–39.

9. Wood-Barcalow NL, Tylka TL, Augustus-Horvath CL. "But I like my body": Positive body image characteristics and a holistic model for young-adult women. *Body Image.* 2010;7(2):106–116.

10. Halliwell E. The impact of thin idealized media images on body satisfaction: Does body appreciation protect women from negative effects? *Body Image.* 2013;10(4):509–514.

11. Hofmeier SM, Runfola CD, Sala M, Gagne DA, Brownley KA, Bulik CM. Body image, aging, and identity in women over 50: The Gender and Body Image (GABI) study. *Journal of Women & Aging.* 2016:1–12.

12. Neff K. Self-compassion. In ed. Mark Leary and Rick Hoyle, *Handbook of Individual Differences in Social Behavior.* New York: Guilford Press, 2009.

13. Andrew R, Tiggemann M, Clark L. Predicting body appreciation in young women: An integrated model of positive body image. *Body Image.* 2016;18:34–42.

Index

About the Author

RENEE ENGELN, PhD, is an award-winning professor of psychology at Northwestern University. Her work has appeared in numerous academic journals and at academic conferences, and she speaks to groups across the country. She is regularly interviewed by the *New York Times*, the *Chicago Tribune*, Today.com, the *Huffington Post*, *Think Progress*, and other national media, as well as local outlets and college student publications. Her TEDx talk at the University of Connecticut has more than 250,000 views on YouTube. She lives in Evanston, Illinois.